❖ *Ending in Ice*

# Ending in Ice

## The Revolutionary Idea and Tragic Expedition
## of ALFRED WEGENER

ROGER M. McCOY

OXFORD
UNIVERSITY PRESS

2006

## OXFORD
### UNIVERSITY PRESS

Oxford University Press, Inc., publishes works that further
Oxford University's objective of excellence
in research, scholarship, and education.

Oxford   New York
Auckland   Cape Town   Dar es Salaam   Hong Kong   Karachi
Kuala Lumpur   Madrid   Melbourne   Mexico City   Nairobi
New Delhi   Shanghai   Taipei   Toronto

With offices in
Argentina   Austria   Brazil   Chile   Czech Republic   France   Greece
Guatemala   Hungary   Italy   Japan   Poland   Portugal   Singapore
South Korea   Switzerland   Thailand   Turkey   Ukraine   Vietnam

Published by Oxford University Press, Inc.
198 Madison Avenue, New York, New York 10016

www.oup.com

Oxford is a registered trademark of Oxford University Press

Library of Congress Cataloging-in-Publication Data
McCoy, Roger M.
Ending in ice : the revolutionary idea and tragic expedition of Alfred Wegener / Roger M. McCoy.
   p.   cm.
Includes bibliographical references and index.
ISBN-13 978-0-19-518857-8
ISBN 0-19-518857-8
1. Wegener, Alfred, 1880–1930. 2. Geologists—Germany—Biography. 3. Grönland-Expedition (1929–1931)
4. Continental drift. 5. Greenland—Description and travel.   I. Title.
QE22.W26.M33 2006
551'.092—dc22     2005023509

9 8 7 6 5 4 3 2 1

Printed in the United States of America
on acid-free paper

*To the creative impulse that induced Alfred Wegener*

*to offer the world new ideas and insights.*

*May that same spirit continue to inspire others.*


*To Sue, wife, best friend, and creative mind.*

*Some say the world will end in fire,*

*Some say in ice.*

—ROBERT FROST

# ❖ *Acknowledgments*

I express my appreciation for the helpful support and assistance from various people in my search for sources and photographs. Jutta Voss-Diestelkamp and Reinhard Krause, archives, and Frank Poppe, public relations, at the Alfred Wegener Institute for Polar and Marine Research, Bremerhaven, Germany, provided many photographs of the Wegener expedition. Ulrich Nickel, historian of the Neuruppin Museum in Neuruppin, Germany, provided family photographs of Wegener and Köppen. Ruben Blaedel of Rhodos Publishers, Copenhagen, gave permission for use of materials from the book *Greenland Ice Cap*. Hubertus Brockhaus, Heinrich Albert Publishing, Wiesbaden, patiently answered many questions concerning obscure copyrights. Peter Hopkins of Kegan Paul, London, provided permission to use quotes from Johannes Georgi's *Mid-Ice*. I especially thank Clare Dudman, a Wegener fan and biographer, for pictures of her visit to Kamarujuk Fjord.

Edward Lueders, University of Utah, and Clifford Mills, Oxford University Press, gave many constructive and welcome suggestions. A special thanks is owed to Lucie Glenn for translations. Sue McCoy read the manuscript at several stages of preparation, and her sharp eye for detail caught many ambiguities and inconsistencies. Her comments were vital and always appreciated.

# ❖ *Contents*

❖ *Ending in Ice*

# 1 ❖

## Scientist and Explorer

*We do not ask for what purpose the birds do sing, for their song is pleasure since they were created for singing. Similarly, we ought not to ask why the human mind troubles to fathom the secrets of the heavens. The diversity of the phenomena of nature is so great, and the treasures hidden in the heavens so rich, precisely in order that the human mind shall never be lacking in fresh nourishment.*

—JOHANNES KEPLER

*Arizona, 2004*

My wife and I visited the Petrified Forest National Park expecting to see some great examples of petrified wood, but not much more. What we found was striking evidence of the movement of continents over great distances. Fossil forests of immense trees along with ferns, crocodiles, and small dinosaurs revealed a past environment of humid tropics similar to the Amazon basin today. Remnants of animals and plants that had lived more than 2 hundred million years ago in a humid tropical environment, but were now lying in this barren Arizona landscape, raised questions that had puzzled climatologists and paleontologists for many years.

Until recently the only accepted answer to the question of how remains of species from the humid tropics were preserved in a desert was that the climate of the region had changed for some unknown reason—possibly by polar wandering. Now the concept of continents moving laterally as part of great crustal plates is so widely accepted that educational displays and materials in national parks adopt it for presentation to the general public.

The story presented at the Petrified Forest National Park explains that the huge trees fell, were covered by silt, mud, and volcanic ash, became cut off from oxygen, and thus did not decompose. Over time silica in the groundwater replaced each cell in the wood, and the logs became petrified. This much is the usual story of petrified wood. However, the story continues. Soon after this forest was buried, the entire supercontinent, named Pangaea, began breaking apart and dispersing over the earth. This great continent had originally included all the major landmasses in one body. Park information shows that what is now the southwest United States had once been located near the equator on the southwestern edge of the supercontinent and had "drifted" from a humid tropical climate into a desert climate.

The German meteorologist and climatologist Alfred Lothar Wegener began researching the concept of continental drift back in 1912 when he presented his first paper on the subject.[1] It seems that Wegener was, in the beginning, simply trying to solve a climatic riddle using evidence from paleontology and geology. The idea of mobile continents has had enormous influence on modern geology, and Wegener was the first scientist who made a major effort to find evidence supporting continental movement.

While Wegener is best remembered today for his hypothesis of continental drift, he was also involved with Arctic expeditions to study the meteorology of Greenland. These research expeditions were made in the early twentieth century when ventures to the Arctic often became struggles for survival.

## United States, 1955

Young geology students in the mid 1950s were exposed to what was then the current theory on Earth's geologic past. Questions about Earth's history concerned the enigmas of mountain building, volcanic activity, crustal subsidence, and rising of the continents. How could one explain the forces needed to crumple a broad area of the earth's rigid crust into tightly folded mountains? Why do mountains often appear in belts along continental margins? Why are volcanoes often associated with the mountain belts? Why do many mountain belts seem to be built at a certain times in Earth's history? Why are there long periods of relative quiet with no mountain building? The basis for the answers to many of these questions in 1950 was that the earth was shrinking and wrinkling due to cooling, rather like a baked potato. This theory had been in vogue for at least a hundred years. However, most professors presented the concepts as facts rather than as possibilities.

To explain the origins of the Appalachian Mountains, for example, early theorists proposed that a landmass, called a borderland, on the east coast of North America had extended an undetermined distance eastward. As this borderland eroded away, all its wasted detritus filled a huge subsiding basin (the Appalachian geosyncline) at the site of the present Appalachian Mountains from Georgia northeastward into Canada. Then, at intervals of millions of years (throughout the Paleozoic era), shrinkage due to contraction compressed the crust into folds that became the Appalachian Mountains.

Like many theories, contraction of the crust raised difficult questions that typically were not discussed with the young students. For example, what became of the oceanic remnant of the offshore portion of the borderlands after they eroded down to sea level? What initiated a period of contractive forces near the end of the Paleozoic era sufficient to cause a compressive motion from the east, pushing crustal rocks into folds to form the Appalachians? Why was the compressive force not more continuous, as one might expect from a steadily cooling body? These questions remained unanswered in 1950.

Another geologic dilemma of that time was the presence of fossils of certain plant and animal species that apparently lived on different continents at the same time. Some leading geologists proposed that land bridges between continents provided migration routes for animals. Proponents cited some examples of land bridges such as the Isthmus of Panama and the Bering Strait, which is not a land bridge today but was known to have been one during Pleistocene periods of glaciation when sea level was lower. Both have served as avenues of migration for humans and animals, and imagining other similar continental connections for migration seemed logical.

The accepted, though unproven, concept of land bridges remained in the minds of most geologists through the 1950s. For example, they knew the *Mesosaurus* was a small reptile with a long slender head that swam in brackish coastal waters, and no one believed it could have crossed the Atlantic Ocean.

5

Rather they accepted that there had been a land bridge along which the *Mesosaurus* had migrated from Africa to South America. Its fossil remains are found in both places today.

Those who supported the idea of land bridges proposed connections from Africa to South America, Africa to India, Asia to Australia, and Europe to North America. This notion persisted even though it raised unanswerable questions. The main question: What became of the land bridges? The usual answer was that the land bridges subsided below the surface of the ocean. When shown that no suboceanic remnant of land bridges exists in the places where they would be expected—as exists in Bering Strait—the response was that the bridges must have subsided through the oceanic crust into Earth's mantle—the layer below the crust. This answer supposes that crustal rocks of low density would sink into deeper rocks of higher density, even though most geologists and geophysicists agreed that would be unlikely. This would be comparable to saying a cork could sink in water. Proponents of land bridges had no better alternative to offer, yet the idea remained at the front of geologic thought and was staunchly defended.

Underlying the thinking about the geologic history of Earth was the belief that "the oceans have always been where we see them now . . . [and] *there has never been general interchange in position between the continental masses and the basins of the ocean* [emphasis in original]." The quote is from a 1933 textbook by Charles Schuchert and Carl Dunbar, both prominent geologists at Yale University in the early twentieth century. Later editions of the book were still in use in the 1950s. Geologists of that day accepted that continents could move vertically, that is up and down, flooding portions of the land from time to time with shallow seas, but they rejected lateral movements of continents.

During the 1950s, in lectures on Earth's history, we heard passing mention of a German named Alfred Wegener who proposed that, rather than being connected by land bridges, the continents had once been joined into a single landmass. In the meantime they had drifted apart to their present positions. This idea was treated as a bit of comic relief in the lectures. Professors did not mention that Wegener had presented extensive supporting evidence for his hypothesis or that international conferences in the 1920s had focused on it. The fact that the drift hypothesis had been generally rejected at those conferences was sufficient to dismiss it from further consideration.

Some of the supporting evidence Wegener had presented was that certain animals living today on different continents were once part of a single group before the landmasses split apart. He cited the manatee, found in tropical rivers in Africa and South America, as a living example of animals that persisted through time after the continents separated. Other extant examples included numerous freshwater fish found in lakes and streams of both Europe and eastern North America and a garden snail found on plants in Germany, the British Isles and Iceland, Greenland, Canada, and the eastern United States. He also mentioned a species of earthworm that is found in Europe and the eastern United States. All these observations he made based on his reading of the literature in various fields. The fact that he was neither a geologist nor a biologist later raised doubts among skeptics about his credibility in making such interpretations.

In Wegener's thinking, the landmasses that formed the bridges then supposedly sank would have been made of continental crustal material and therefore could not have sunk into the denser oceanic crust and upper mantle. He stated that the additional landmass from the land bridges would have raised the sea level and created marine environments where terrestrial or coastal environments were known to have been. The only choice in his mind was between permanence or drift, immobility or mobility, and he felt the biological and geological facts favoring mobility were overwhelming.

Only a few mobilists (those believing in lateral movements) accepted the possibility of continental drift. Most earth scientists—especially in North America—were fixists, rejecting the possibility of lateral continental movement. Wegener's revolutionary mobilist hypothesis was reduced to no more than a quaint footnote to the geologic ideas of the mid-twentieth century, and a handy joke for class lectures.

## Germany, 1880–1928

Nothing about Alfred Wegener's education prepared him for research on continental drift. He had no background in geology or geophysics. His associates included primarily astronomers and meteorologists. Professional interests led him to pioneering efforts in meteorology in remote Greenland. His research followed two tracks that appear to have little convergence:

gathering first-time atmospheric data over the Greenland ice cap and proving that the continents moved. He attained widespread recognition, or notoriety, in each area. What would explain the course of his life? What could have drawn him into such an apparent professional diversion as continental drift for which he seemed to be untrained? Was his research in continental drift in conflict with his main interests, or was it perfectly reasonable for him to pursue such a radical hypothesis in the face of strong opposition from geologists?

His father, Richard, who had a Ph.D. in theology and classical languages, taught in a highly regarded secondary school in Berlin. Alfred's mother, Anna Schwarz, came from the village of Zechlinerhütte about fifty miles north of Berlin. Richard and Anna Wegener had five children; the youngest, Alfred Lothar, was born in Berlin in 1880. Only Alfred, his brother Kurt and sister Tony lived to adulthood. Figure 1.1 shows Alfred at age forty-nine.

Having no interest in his father's area of study, Alfred followed his own interest in science. He entered Friedrich Wilhelm University in Berlin to study mathematics and natural science with an emphasis in astronomy, completing his

FIGURE 1.1
Alfred Wegener shortly before his fourth and final expedition to Greenland in 1930.
(Used with permission of the Neuruppin Museum, August-Bebel Strasse 14-15, Neuruppin, Germany.)

studies there in 1902, at the age of twenty-two. Diploma in hand, he went to work as an astronomer in the Urania Society observatory in Berlin and conducted research toward his Ph.D. in astronomy. In 1904 he passed his dissertation examination magna cum laude. His topic was a modernization of thirteenth-century Alfonsine astronomical tables for determining orbits of Earth's moon and the five known planets.

Wegener wanted to shift his career away from astronomy because it did not offer sufficient opportunity for travel and physical activity—he wanted some adventure in his life. Also it became clear to him that he would need more mathematics to advance in astronomy, and he felt he had marginal ability in that area. He chose meteorology.

In 1905 Wegener took a job with his older brother, Kurt, also a meteorologist, as a technical aide at the Aeronautic Observatory in Lindenburg, Germany, outside Berlin. There they learned modern techniques for atmospheric research using kites and weather balloons, which Alfred later used in Greenland. He and Kurt also made ascents in hot air balloons and became active recreational balloonists. In 1906 they exceeded a world record by seventeen hours when they stayed aloft for fifty-two hours during a balloon excursion from central Germany to northern Denmark and back into Germany near Frankfurt. At age twenty-six, Alfred's taste for adventure was evident in his record-setting balloon ride and an upcoming expedition to Greenland.

Wegener's meteorological experience brought him an invitation to join the *Danmark* expedition led by Ludvig Mylius-Erichsen in 1906. Mylius-Erichsen studied the Eskimos of Greenland, and Wegener served as the meteorologist. On this expedition, Wegener, drawing on his experience with weather kites, became the first person to make use of meteorological kites and balloons for collecting high-altitude data in a polar climate.

In preparation for the trip Wegener sought advice from Wladimir Köppen, who was head of the meteorological kite station at Grossborstel near Hamburg. The meeting with Köppen led to a long and fruitful relationship. At first Köppen was a mentor to Wegener, but he soon perceived the young Wegener as a worthy colleague and the relationship became one of research collaborators.

During the 1906 expedition, Wegener learned about surviving in the Arctic, and how quickly mistakes can become deadly. The expedition leader,

Mylius-Erichsen, and two companions died while on an extended winter excursion away from base camp because they ran out of food before they could make it back to the coastal camp. Despite this tragedy, the rest of the expedition stayed on through the planned two years, completing their research in 1908. Wegener learned that even the most meticulous planning is at the mercy of the Arctic environment.

In 1909 Wegener found a teaching job at the University of Marburg. As a *Privatdozent*, Wegener did not have a regular salaried faculty position, rather his pay came from fees students paid to attend his lectures plus occasional honoraria for lecturing to special groups. Financially this was a very difficult arrangement for him, but it allowed him to work toward qualification for a professorship at a university.

Wegener's lectures were well attended, and he had a reputation for making complex ideas easily understood. He taught courses in astronomic position finding, physics of the atmosphere, atmospheric optics, and general astronomy. Although he had chosen to pursue meteorology, and had been hired on the basis of his publications on weather kites and balloons and his meteorological experience in Greenland, he could not escape astronomy. While at Marburg, he wrote and published his first book, *Thermodynamic der Atmosphäre* (1911). This soon became the standard textbook for atmospheric physics in Germany.

Alfred first met Wladimir Köppen's daughter, Else, after he returned from Greenland in 1908. He became a regular and welcome visitor to the Köppen home. A spark of interest could have struck between Alfred and Else during these first visits, even though she was only sixteen—twelve years younger than Wegener. Else was once excused from school to attend a meeting where Wegener was speaking of his Greenland expedition. She wrote of the occasion, "I was sixteen years old at the time and enthusiastic about everything extraordinary. What beautiful slides Dr. Wegener showed and how clearly he spoke! He was still tanned from the Arctic sun and the sea air. His gray-blue eyes beamed light from his dark face."[2]

She was smitten with this twenty-eight-year-old Arctic explorer. Albert must have felt drawn to Else as well. She wrote that Albert would often stop in the middle of his dinner conversation and give her a big smile or tease her a bit. A smile may have been unusual for Wegener, as he is described by some colleagues as more often serious than smiling. Though not striking in appearance, his slender, muscular build, prominent forehead, and penetrating blue eyes made a lasting impression on young Else.

Else described the friendly, warm feeling after dinner when the guests would sing "old meteorologist songs."[3] Singing at informal gatherings was popular in the early twentieth century. Although the meteorologist songs they sang are not identified, one of them may have been based on a poem by Johann W. Goethe, the famous German writer, poet, and philosopher of the late eighteenth and early nineteenth centuries. Goethe took a strong interest in many scientific subjects, including climatology. When the Englishman Luke Howard published a classification of clouds in 1815, Goethe felt inspired to write a cloud poem in Howard's honor.[4] The poem has lengthy praises of the beauty and majesty of cirrus, stratus, cumulus, nimbus clouds, and the atmosphere in general. Every educated person in Germany knew Goethe's works well, and this poem could have been set to music as a song for meteorologists.

Visitors at the Köppen household included guests from various disciplines, and Else reported that congenial dinner conversations covered many subjects of scientific interest. Although she does not record the exact subject of these conversations, it is likely that Köppen would be interested in showing visitors his work and discussing ideas for a possible project. Probably a study of paleoclimates, which he eventually published with Wegener, had already begun to germinate in Köppen's mind.

Paleoclimate discussions with Wegener and other visitors would have focused on questions of geologic and paleontologic evidence that might indicate the climates of the past. Where had past glaciation occurred? Where can evidence for past deserts, such as petrified dunes and evaporite deposits of salt and gypsum, be found? Where can coal deposits indicating wet tropical environments be found? Wegener's visits to the Köppen home led him to comment that this time had been "an almost inexhaustible source of stimulation."[5] These paleoclimate questions could have suggested to Wegener the idea of shifting continents in order to make a better fit of climate indicators.

Wegener made a second trip to Greenland (1912–1913) with a Danish expedition led by Johan Peter Koch, a member of the Mylius-Erichsen expedition. Wegener took along one of his most promising young students, Johannes Georgi, who became a primary researcher on Wegener's later Greenland trips.

During Wegener's absence, Marburg University held his teaching position for him.

This second trip to Greenland gave Wegener some valuable experience. Here he nurtured his ambition to study the climate of Greenland—a virtually unknown subject. Wegener and three others stayed on the east coast of Greenland over the winter collecting meteorological data, then in April trekked with Icelandic ponies across Greenland. They had an unexpectedly hard trip and narrowly averted disaster. Their food supply was dangerously low when some Eskimos found them and helped them reach a small settlement on the west coast in mid-July. None of the ponies survived the trip, although the party tried to save the last one, weakened and starving, by pulling it on a sled. Though the experience with ponies was unsuccessful, Wegener thought there would be a great advantage to a combination of ponies and dogs.

Before Wegener left for Greenland in 1912, he proposed to Else Köppen. By then Else was twenty and Alfred thirty-two years old. During Alfred's year in Greenland, Else went to live in Oslo with the family of Vilhelm Bjerknes, a prominent Norwegian meteorologist. There she taught German to the family's children while she learned Norwegian and Danish. Else was later able to translate reports from the Danish expedition into German. When Alfred returned from Greenland in 1913, he and Else were married, and began seventeen years of a happy marriage. Their first child, Hilde, was born in 1914, the same year that World War I erupted. Figure 1.2 shows Alfred and Else in their home at Marburg.

Wegener, an officer in the German army reserve, was immediately called to active duty in Belgium. Else wrote of the irony of her husband engaged in war. "How often he must suffer under the brutality of these mass murders. He is obligated to lead his men against the enemy, the 'enemy' with whom he recently, perhaps, engaged in an exchange of scientific ideas."[6]

Wegener was wounded in the shoulder during his first experience with combat and was sent home on leave to recuperate. He arrived home in the summer of 1914, when his daughter Hilde was only three days old, stayed for only two weeks, and returned to the trenches on the western front. A few weeks later he received a more serious wound in the neck. After being treated for his second injury, he was once again sent home for a longer recuperative leave that lasted until April of 1914. During his recuperation he continued his research and writing about continental drift, and the first edition of his book

FIGURE I.2
Else and Alfred Wegener at home in Marburg, 1913. (Used with
permission of the Neuruppin Museum, August-Bebel Strasse
14-15, Neuruppin, Germany.)

*The Origin of Continents and Oceans* (*Die Entstehung der Kontinente und Ozeane*)
was published in 1915.

Wegener resumed active duty at Mülhausen, Germany, where he served as
a military meteorologist, remaining at his post until the summer of 1917. This
was an ideal arrangement for Alfred and Else. She traveled to Mülhausen frequently during this period, and the couple was never separated for longer
than three months at a time. In the summer of 1917, however, the army first
moved Alfred to Sofia, Bulgaria, and then later to Estonia, where he served as
a weather observer. During that time he completed some new research on

whirlwinds and waterspouts. In 1918 Alfred returned to Marburg and the university. His income was meager, and food and fuel were scarce. The outlook was bleak. That same year another daughter, Käthe, was born.

Within a year, Wegener was chosen to succeed his retiring father-in-law, Wladimir Köppen, as head of the department of theoretical meteorology in the national weather service at the German Marine Observatory in Hamburg. Alfred and Else moved into the ground level of Köppen's house, and the Köppens moved to the second story. The house had a garden with fruit trees, berry bushes, and a play area for the children—a welcome change after years of war and privation.

In 1919, Alfred's brother Kurt also became a department head in the same observatory, and the brothers worked together for the next five years (1919–24). In addition to his work as a meteorologist at the observatory, Alfred gave lectures at the newly established University of Hamburg. While in Hamburg he published research on lunar craters, proposing that they were impact craters rather than the result of volcanic action. The origin of lunar craters was a controversial issue at that time, but Wegener's view eventually became the accepted one.[7] The Wegeners' third and last child, Charlotte, was born in 1920 during this time in Hamburg.

In 1924 the University of Graz, Austria, offered Wegener the professorship he had long sought. Other universities had passed him over because they thought his interests were too diverse and that he tended to be away on expeditions too much. Although these traits might have been attractive to universities considering that all his work, especially the expeditions, gained enthusiastic public attention, some universities, conscious of their academic image, may have thought the continental drift hypothesis was too controversial. The University of Graz did not.

Wegener's expeditions and his drift hypothesis had made him a prominent scientist throughout the German-speaking area of Europe. He moved his family to Graz and became a professor in the department of geophysics and meteorology. Soon the family became Austrian citizens. Else's biography of Wegener describes this period as their happiest years. Wladimir Köppen, by now retired, and his wife, Marie, moved to Graz and lived with Alfred and Else. A 1930 photograph shows the family group while in Graz (figure 1.3). Wladimir Köppen was not in the picture—perhaps he was behind the camera. Wladimir is shown with Marie in figure 1.4.

FIGURE 1.3
Alfred Wegener, Else Wegener,
Marie Köppen, and Kurt
Wegener in Graz, Austria, 1930.
(Used with permission of
the Neuruppin Museum,
August-Bebel Strasse 14-15,
Neuruppin, Germany.)

FIGURE 1.4
Marie Köppen and Wladimir
Köppen in Graz, 1933.
(Used with permission of the
Neuruppin Museum,
August-Bebel Strasse 14-15,
Neuruppin, Germany.)

Alfred Wegener's professional training was in meteorology and climatology, but his work on continental drift included a great deal of geology. Although some reference sources today refer to him as a geophysicist or geologist, Wegener would probably not have identified himself in that way. Wegener and his drift hypothesis were scorned in part because he was thought to be working outside his field of expertise. Almost certainly, his interest in the geologic aspects of continental drift was fueled by a desire to find a solution to paleoclimate questions. But he had committed the unpardonable error of treading on the wrong turf.

# 2 ❖

# Wegener's Shocking Idea

*That is the essence of science: ask an impertinent question, and you
are on the way to a pertinent answer.*

—JACOB BRONOWSKI

Alfred Wegener was not the first scientist to imagine the possibility that
continents could move, but he was the first one to compile extensive
geological, geodetic, and paleontological evidence. His evidence had many
gaps and ambiguities, partly because of incomplete world maps of geology
and fossils at that time; but the great amount of supporting evidence he compiled could not be easily ignored by geologists. So they did not ignore it. Instead, the earth science community convened conferences and attacked both
Wegener's theory and him personally. Only a handful of geologists in Europe
and South Africa thought Wegener's ideas deserved further consideration.

The pattern of congruence of the west coast of Africa with the east coast
of South America is a visual clue to the theory of continental drift. Francis
Bacon is often mentioned as the first writer to point out the coincident pattern of continents as a matter of passing interest in 1620. That was soon after
reasonably accurate maps of continental coastlines first became available, so no
one could have written on the subject much earlier. In the early nineteenth
century the famous German geographer and explorer Alexander von Humboldt also commented on the similarity of mountain ranges in Brazil and west
Africa but did not suggest that they had moved apart. Rather von Humboldt

thought that some land was removed by catastrophic erosive action that cut a deep valley, which became the Atlantic Ocean. This notion was pure speculation based on no evidence.

In 1858 Antonio Snider-Pelligrini published a book *La création et ses mystéres dévoilés* (*Creation and Its Secrets Unveiled*). In it he showed a map with both South America and North America positioned next to Africa. Snider-Pelligrini did not pursue further development of the idea that continents had separated, except for that one map. He proposed a catastrophic event that caused the continents to separate. As Wegener began to publish, he gave no indication that he had seen Snider-Pelligrini's book. But in later revisions, Wegener included some historical background regarding his predecessors, including Snider-Pelligrini. Critics fault Wegener for not acknowledging this source from the beginning, implying that the idea must have come to Wegener through Snider-Pelligrini's book.

One recurring criticism was that Wegener was unqualified to develop geologic theory. What did motivate Alfred Wegener, a meteorologist, to become so intrigued with the idea of continents moving? His background and experience are focused entirely on meteorology and astronomy. His expeditions to Greenland centered on applications of meteorology and glaciology. What drew him to continental drift? The answer could lie, in part, with his mentor, collaborator, and father-in-law, Wladimir Köppen.

Köppen produced his first climate classification scheme in 1900. This effort was based partly on an existing vegetation map of the world made in 1867. In 1918 Köppen improved his methods to include rainfall and temperature as criteria for climate classification. His famous wall map of world climates based on this approach first appeared in 1928. The classification system and the map resulting from that work are still in widespread use around the world with few modifications.

After his first work on world climates Köppen began to turn his interest to the climates of the geologic past. During visits to Köppen's home, Wegener and Köppen would have talked about theories of climates, both current and past. They found their interests complementary and agreed to collaborate on the study of past climates. They could have begun discussing and researching the subject as early as 1910, after Wegener returned from his first expedition. However, their book *Die Klimate der Geologischen Vorzeit* (*Climates of the Geologic Past*) did not appear until 1924.

In order to study past climates Köppen and Wegener had to rely on existing information about the locations of fossil nonmarine plants and animals. The clear relationship between plants and climates was the basis for climatic boundaries on Köppen's map of present-day climates. In the same manner, fossil plants were used to deduce the past climate in which the now fossilized plants once grew. One point of interest is the fossil plants found in coal beds. Knowing that coal forms from dense vegetation in marshy, tropical or subtropical environments provides good information about the climate existing at the time the vegetation grew. Also the type of fossil plant (palms or ferns, for example) provides important information when we know the climate in which similar plants grow today. Fossils of animals may be less indicative of climate than plants, but they can indicate whether the animal lived on land, freshwater near land, or in an ocean. The fossil of an animal found in sediments derived only from land or freshwater can be presumed to have lived in a nonmarine environment. When the same fossil is found in another part of the world, a similar nonmarine environment can be assumed to have existed in that location.

Wegener mentioned finding a reference book on the known fossils of the world in 1911. As he studied the book, he became intrigued by the similarities of nonmarine fossils of Paleozoic strata in Africa, Brazil, and India. This book provided much information on the distribution of fossils and allowed Köppen and Wegener to begin mapping past climates.

Moving from Wegener's discussions of paleoclimate in Köppen's home to an insight on continental drift would be an easy step for a person adept at making connections. Evidence for Paleozoic glaciation in such diverse locations as southern Africa, South America, and India could be explained as one contiguous event simply by moving the continents back to a single land mass. It would explain simultaneous glaciation in areas that are now widely separated and tropical.

Wegener mentions his excitement when looking at a friend's new atlas. He wrote Else in 1911, "For hours we examined and admired the magnificent maps. At that point a thought came to me. Does not the east coast of South America fit exactly to the west coast of Africa, as if they had been connected in earlier times?"[1] Certainly Wegener must have looked at maps many times and noticed those distinctive jigsaw shapes, even as a child. However, this time he had a more insightful view, which probably emerged from his stimulating discussions in Köppen's home about paleoclimates and the possible explanations for distributions of past glaciation, deserts, and coal deposits.

Pushing the continents back together provided Wegener an alternative to the land bridge concept. If all the continents were formed into one large landmass, patterns of fossils, geology, and climates on separate continents became reconnected, and it made sense to him that those continents must have been joined. For Wegener this was a tremendous breakthrough—the flash of insight—the great ah ha! This germ of an idea grabbed him and drove him to research the question of continental drift in greater detail than had anyone ever had before.

Wegener began to search for other information to support his idea. He first began thinking in terms of continental drift around 1910. Within four months he made his first oral presentation on his drift hypothesis to a geological society in Frankfurt am Main. Four nights later he presented a similar paper in Marburg. After this he was fully committed to a complete development of the drift hypothesis with no turning back and no hesitation.

After seeing these initial papers severely criticized, Köppen cautioned Wegener that he should not expand into other fields for there was enough yet to be done in his own field. Wegener wrote in reply, "If it can be proven that now sense and reason come into the entire geological development history of the globe, why should we hesitate to throw the old perception overboard? Why should one hold back for ten or even thirty years with this idea? I do not believe the old concepts will last ten more years."[2] It seems apparent that by this time Wegener had made a transition from working on a climatology question to working on a geologic one.

In 1910 Frank B. Taylor, an American geologist, presented a proposal for continental drift saying Africa and South America had been joined before separating along the Mid-Atlantic Ridge, and that they had moved away from the ridge in opposite directions.[3] He proposed that large land masses over the north and south poles had crept toward the equator, and their forward edges were folded into mountain ranges, explaining the prevalence of coastal mountain ranges on several continents. The gaps left by the moving continents became the Arctic, Atlantic, and Indian Ocean basins. He considered Antarctica to be a large fragment that never moved away from a polar position. Taylor proposed that the cause of the continental movement was a combination of tidal action and an increase in the rate of Earth's rotation, forcing landmasses to move away from the poles by centrifugal force. In Taylor's hypothesis the increase of rotation was caused by the capture of a comet which became the moon. He assumed this occurred as recently as the Cretaceous period (65–144 million years ago). To his credit, Taylor's hypothesis included the idea of crustal sheets moving as a unit.[4]

Taylor thought tides must have become much stronger in the Cretaceous period because the moon had been trapped in an elliptical orbit at that time. At the perigee position of the lunar orbit the moon would have been only 38,000 kilometers (about 24,000 miles) from the earth according to Taylor. This idea raised a question. If lunar tides did not exist before the Cretaceous, then what forces would account for mountains formed in the Paleozoic? Taylor's answer was that solar tides were much stronger in pre-Cretaceous time because the earth was much closer to the sun. Would not climatic evidence be present if the earth had been closer to the sun? Would Paleozoic glaciation have occurred if the earth had been closer to the sun? He provided no answer to these questions. The extensive speculation without evidence prevented Taylor's hypothesis from leading to serious debate.[5]

Taylor's ideas might have attracted more attention had not Wegener's hypothesis appeared so soon afterward. Taylor made no attempt to reassemble the continents to their former position, except to say that Greenland had been attached to North America. His hypothesis that the moon was captured so late in Earth's history gave the proposal a sense of catastrophism that was not generally acceptable to the proponents of uniformitarianism—the idea that geologic processes of the past are the same as those today.

Because he had presented the notion of continental mobility, Taylor was at first considered a co-originator of continental drift, particularly in North America, and Americans called it the Taylor-Wegener hypothesis. Eventually Taylor's name was dropped as the greater extent of Wegener's work became apparent.

Wegener's hypothesis, unlike Taylor's, included a synthesis of converging evidence from geology, geophysics, paleontology, and geodesy (the science of measuring the earth) in support of his proposal. He acknowledged that some aspects of Taylor's theory were similar to his own, but asserted that Taylor had given continental drift a very cursory explanation. Wegener said that he became acquainted with Taylor's works only after his own theory in its main outlines had already been worked out.

Wegener always maintained that he had not known of Taylor's theory until later. Yet Taylor's paper, published in 1910, had been read at a Geological Society of America meeting in 1908. A review, by Jesse Hyde, of Taylor's paper appeared in the German publication *Geologische Zentralblatt* in 1911. Most scientists of the day doubted that Wegener could have failed to notice all these papers.[6] Taylor claimed in a 1931 paper that Wegener had, in 1911, published a brief review, only twenty lines long, of Taylor's 1910 paper. Taylor, however, was unable to find a copy of Wegener's review, nor could he recall a citation for it. Nor did anyone else ever find such a review.

History of science shows other examples of new ideas appearing almost simultaneously from independent sources. When this happens there is often reluctance among the discoverers to share the recognition. Each wants to play a solo. This situation existed even before the days of research secrecy and competition created by grants and Nobel prizes. One widely known example is the simultaneous development of the theory of evolution through natural selection by Wallace and Darwin. Though Wallace published first, and Darwin openly lifted some of his key ideas, Darwin is the one who publicized his ideas and received all the recognition and fame. Wegener's case vis-à-vis Taylor is stronger because he produced more thorough evidence. He would have received the same recognition even if he had said that part of his idea came from Taylor.

In 1911 Howard Baker also wrote of continents moving due to planetary perturbations. He proposed that Venus and Earth came into close proximity and caused the moon to be ripped from the earth. The resulting chasm became the Pacific Ocean and caused the other landmasses to split apart along the Mid-Atlantic Ridge, forming the Atlantic Ocean.

Other early publications on the idea of continental movement never provided a body of supporting evidence such as Wegener provided, so he rightly became known as the originator of the theory of continental drift. Some authors have called him the Father of Continental Drift. He was, in fact, the only one who developed a comprehensive hypothesis with convergence of several lines of evidence on a global scope. His use of coastal outlines was only a starting point.

## The Continental Drift Hypothesis

Motivated by paleoclimate questions, Wegener began the accumulation of data to support his hypothesis that continents had moved laterally. Initially, in 1912 Wegener matched only the present-day coastlines. Then he expanded the continents to a size they might have had before the folding of mountains shortened them by several hundred miles. Later he tried matching the continents along the edge of the continental shelf. This last effort provided the best fit of landmasses because it was independent of sea level fluctuations that had occurred since the continents drifted apart.

Another element in the hypothesis was based on the fossil similarities of west Africa and Brazil during the late Paleozoic era (more than 250 million years ago). The idea that the continents were joined at that time began to make sense to him. He gathered all information—geologic, paleontologic, geodetic—that might support the notion of continental drift.

His book *The Origin of Continents and Oceans* caused a stir in Europe when it was published in 1915, and again when a second edition appeared with additional material in 1920. The third edition came in 1922 and a fourth by 1929. The third edition was translated into French, Spanish, and English in 1924. Once the English version became available, interest from North America and Britain increased remarkably, and the basic tenets of the continental drift hypothesis became widely known. The fourth edition was reprinted as a fifth edition in 1936 and as the sixth edition in 1941 and 1962. Another English version became available in 1966, translated from the fourth edition. By the fourth edition Wegener's book had grown from 94 to 231 pages because of his additional research, and it could be read in seven languages. Few books in the natural sciences had ever been so widely read or provoked so much international discussion.

The concept of a two-layer crust on the earth was of particular importance to the Wegener hypothesis. He felt that the lower density of the continental blocks allows continents to ride above the denser rocks of the ocean basins. The oceanic crustal material forms a continuous, relatively thin cover over the mantle (the layer under the oceanic crust). The continents form distinct, separate masses that are much thicker and appear to "float" on the oceanic crust. Only about one-third of the earth is covered with continental crust.[7]

Wegener said that all the continental masses once formed a single large landmass, which he called Pangaea, until the lower Jurassic period (200 million years ago). Then Pangaea began to break apart along rift lines, such as the Mid-Atlantic Ridge. He fit the pieces together and found that a number of features (mountain ranges, sedimentary rock formations, lava flows, glacial deposits, and the distribution of fossil flora and fauna) matched on different continents.

Successful fitting of continental coastlines requires more than matching pieces. Like a jigsaw puzzle, the picture must make sense too. Wegener recognized that geologic features and fossil locations on his rejoined continents must match in order to be convincing. For that purpose he sketched in structural features that appeared to start on one continent and continue on another continent across the Atlantic. He showed the Appalachian Mountains extending to the British Isles and northern Europe. The Sierra de la Ventana near Buenos Aires matched the Cape Mountains of southern Africa.

Wegener included several other geologic features that he felt matched perfectly. The Zwartberg folded range in South Africa lines up with the Sierras of the Buenos Aires Province. The plateaus of Africa and South America have matching sequences of rocks that include uncommon varieties of igneous intrusives and kimberlite pipes with diamond fields in both continents. The Paleozoic system of Karoo sediments in South Africa is matched by the Santa Caterina formation in Brazil. Both contain *Glossopteris* (a fern) and *Mesosaurus* (a freshwater reptile) fossils. Wegener showed that the Paleozoic folded ranges of Brittany and southwestern Ireland continue in Nova Scotia and southeast Newfoundland, making the coal fields of Belgium and the British Isles align with coal fields of the Appalachians. Thick Devonian Old Red Sandstone forms a band extending from the Baltic States, southern Norway, and the British Isles to Greenland and North America. The Caledonian folded mountains of Scotland and Ireland continue in Newfoundland. The rocks of the mountain systems of Algonkian age in the Hebrides and northeast Scotland continue in Labrador. He showed that the terminal glacial moraines of the Pleistocene ice sheets in Europe unite with those of North America, although they are now separated by 2,500 kilometers (1,550 miles) of ocean and the American glacial moraines are 4.5 degrees of latitude south of the European moraines. Many of these features that Wegener matched were disputed by his critics, especially the matching of the glacial moraines of North America and Europe. A more recent map of the southern portion of

Pangaea (Gondwanaland) shows the distributions of four fossils that were living in the Paleozoic before the continents began to rift apart (figure 2.1).

Glacial tillites (consolidated glacial debris) on the Falkland Islands were of particular interest because they contain numerous erratic boulders of igneous and metamorphic rocks that are foreign to the islands. Wegener was convinced that boulders in southern Africa could be found in the Brazilian tillite, and that diamonds of South Africa, Brazil, and India could have a common source. He had no substantiated observations or research to support this, but the logic seemed clear to him.

To Eduard Suess (1831–1914), an Austrian geologist, the similarity of glacial tills and geologic structures dating to the Paleozoic (more than 245 million years ago) meant that ice sheets, basalts, and sediments had once been continuous over immense areas of the globe. Wegener visualized a smaller area

FIGURE 2.1

Distribution of four fossils found in continents of the southern hemisphere. *Mesosaurus* and *Glossopteris* are from the Paleozoic; *Lystrosaurus* and *Cynognanthus* are from the early Mesozoic. (Adapted from Kious and Tilling, 1996.)

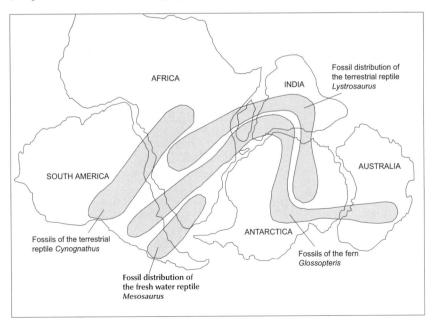

of ice coverage by fitting continents together and assuming later separation by continental drift.

Wegener drew the analogy of fitting together a torn newspaper and finding the lines of print make sense again. If two segments of even one geologic feature matched across the Atlantic like lines of print, the odds would be ten to one in favor of his theory. Matching multiple independent features raised the odds to 1 million to one in favor of continental movement. He conceded that his statement of the odds could be somewhat exaggerated, but nevertheless felt his case for drift was beyond doubt.

Wegener found no features in Spain and northern Africa that match with North America, and concluded there must have been a piece of open ocean at that location while most of those continents were in contact. He identified the Atlantic islands, the Azores, the Canaries, and Madeira as sloughed remnants left in the wake of separating continents.

Wegener wrote that continental rift valleys marked the beginning of a breakup of continents. Then the rift valleys gradually widened into oceans as the continental blocks plowed through the oceanic crust. The resistance the continental blocks met as they moved caused the leading edges, such as western sides of North America and South America, to crumple into mountain ranges. Mountain building that started at the end of the Cretaceous period and continued into the Tertiary period could be accounted for in this manner. Examples he gave of mountains formed in this way included the Himalayas, as India shoved into Asia, and the Andes and Sierra Nevada Mountains, as South America and North America pushed through the resistant oceanic crust. Wegener found this explanation of mountain building made more sense than the widely accepted idea of the earth shrinking due to cooling. The cooling and shrinking explanation for mountains left a number of unanswered questions about the uneven distribution of mountains and the irregular timing of mountain building episodes. Objects that wrinkle from shrinkage tend to wrinkle simultaneously and uniformly over the surface.

Wegener's hypothesis explained islands and arcs of islands, such as those in the western Pacific, as remnants left behind by moving continents. He proposed that trenches in the western Pacific Ocean floor were caused by tension fractures torn open as Asia moved to the west. He believed that trenches of the eastern Pacific resulted from the ocean floor's dipping into a deep sag beneath oncoming continents. Wegener's explanations of features such as mountains, islands, and ocean trenches were all radical for the time. A few of his interpretations have been substantiated over the years, but many have been discarded as research revealed new data.

According to Wegener the South Atlantic opened in the late Mesozoic (around 66 million years ago) and continued separating longer than the North Atlantic, which began opening at the end of the Pleistocene as the ice caps were melting (about 11,000 years ago). North America split away westward and 4.5 degrees of latitude southward. His notion of the opening of the North Atlantic as recently as 11,000 years ago was greeted with much skepticism: continents would have to have moved more than 183 meters (600 feet) per year to reach their present position in so short a time span.

Wegener needed to identify a force strong enough to move continents against the resistance of the underlying oceanic crust. Otherwise, his hypothesis had no chance of being accepted by the international geologic community. Some of his supporters argued against outright rejection of the concept of continental movement even without identifying a credible force for movement. Doubters said that without a credible force to explain continental movement, there was no basis for acceptance of the hypothesis.

Wegener proposed two forces. One was centrifugal force, which would impel a continental mass toward the equator. He based this on work done a few years earlier by a Hungarian physicist named Eötvös who had calculated that the earth's spinning should create a small force tending to move the continents away from polar regions toward the equator. The second force was the tidal attraction of the sun and moon exerting a drag on continental blocks, slowing Earth's rotation, and moving the continents westward relative to the underlying layers. Wegener recognized that these forces were very small relative to the size of the task, but he explained that when applied for long periods of time, they would become effective enough to move continental blocks through the oceanic crust, which would spread open as the continent approached and close behind it as it passed. Many geophysicists agreed that both forces exist but rejected the notion that they would be strong enough to move continents or fold mountains. Only by the fourth edition of his book did Wegener modify his concept to include Ampferer's (also Schwinner's and Kirsch's) idea of slow convection currents in the earth's mantle layer causing continental blocks to ride along on oceanic crust that also moved.

As early as 1906 the director of the Austrian Federal Geological Institute,

Otto Ampferer, proposed the possibility of currents in the viscous material of Earth's mantle underlying the crust. When Wegener first published his drift hypothesis (1915), Ampferer rejected the external forces that Wegener suggested and wrote that currents in the mantle must be the moving force. He drew a schematic illustration that was close to modern concepts of plate movements. Another Austrian, Robert Schwinner from Graz University, also thought convection currents might exist in the mantle. This idea was not accepted by most geologists and geophysicists, who said that seismic data showed that the mantle was too rigid for currents to exist within it.

Wegener made no mention of communication or familiarity with either Ampferer or Schwinner. This is surprising considering that all were in the same country and Schwinner was in the same university as Wegener. Schwinner wrote a textbook containing a section on continental drift with attention to its drawbacks. This might have contributed to Wegener's lack of contact with Schwinner.

In his third edition (1924), Wegener mentioned ideas by both Ampferer and Schwinner and acknowledged the possibility of a role for convection in the mantle as a moving force. In this later edition Wegener also mentioned gravity as a force that could be moving continents. He proposed that buildup of heat may cause a bulge under continental blocks, while heat escapes more readily through the thinner oceanic crusts. As a result, the continents would move by gravity downslope from the crest of the bulge like enormous slow landslides. This suggested that Pangaea broke apart and slid in opposite directions down the slopes of a bulge. North America and South America slid westward, and Eurasia slid eastward all forming mountains on their leading edges.

The fourth edition (1929) included the idea that the position of Earth's polar axis of rotation may have changed since Permian time (245 million years ago) and redirected the centrifugal force effect. The book on paleoclimates by Köppen and Wegener (1924) also included axis relocation as one possible explanation of climate changes.

Wegener had made measurements of latitude and longitude at repeated intervals as precisely as possible at that time. Surveying various stations on an island off Greenland and comparing those data with measurements made in 1823 by another explorer, he concluded that Greenland was still moving westward. Later examination showed that Wegener's margin of error in his computations was greater than the calculated amount of movement.

Three measurements of latitude and longitude at locations on Sabine Island near Greenland were available to Wegener: Edward Sabine's from 1823, Borgen and Copeland's from 1870, and J. P. Koch's from 1907. The three surveyors measured different sites of Sabine Island, but Wegener tied them all to Sabine's point by triangulation. All three of these measurements had used lunar charts for measuring longitude. Lunar charts were an accepted method for navigation if the objective was simply for a ship to find home port, but the potential margin of error was too great for the precise measurement needed to detect small continental movements.

Wegener calculated that the mean error in the three measurements using lunar charts was 124 meters (407 feet). But his result showed that Greenland had moved westward relative to Greenwich by 420 meters (1,378 feet), or 9 meters (30 feet) per year between 1823 and 1870. Between 1870 and 1907 Wegener calculated a total movement of 1,190 meters (3,904 feet), or 32 meters (105 feet) per year. These rates of movement suggested that continents could have moved around the circumference of the earth in only 1 million to 5 million years. With such large movements, it is little wonder that his margin of error seemed insignificant. He could not have realized that his margin of error was three thousand times larger than the actual movements measured today. Today's calculations of continental movement average about 4 centimeters (1.6 inches) per year.

The use of radio time transmissions for determinations of longitude began in 1922. A radio station was built in western Greenland for the purpose of receiving radio time transmissions, and measurements of longitude were made at five-year intervals. In 1927 the measurement indicated a westward drift of 36 meters (118 feet) per year. Wegener felt that this proved that a displacement of Greenland was still in progress. None of these measurements held up under more precise time measurements in later years.

The main skeptics of Wegener's hypothesis were geologists and geophysicists, but scientists such as climatologists, botanists, and paleobotanists soon began to support the idea. The drift hypothesis answered some of their most puzzling questions. Instead of relying on radical climate changes to explain past distributions of plants, animals and glaciers, drift supposed that continents simply moved from one climate zone to another, carrying their life forms and geologic formations with them. Thus the inhabitants of the earlier zone were carried into the new zone where they either survived or died out and became fossils. The building of mountain ranges, such as the Sierra Nevada in the western United States, could account for climate change within the continent. As the mountains arose they blocked the flow of moisture into the continental interior, creating a dry climate where a wetter climate had been. The same changing situation could be shown affecting climates in Tibet north of the Himalayas and east of the Andes in Argentina—mountains formed by continental drift.

For glaciologists the drift hypothesis was appealing as an explanation of upper Paleozoic era glaciation (240 million to 300 million years ago). In the current warm climates of southern Africa, India, and South America, glacial moraines and striations of rocks were found with fossils of plants typical of cool climates. This phenomenon could not be explained by present-day locations of continents in tropical climate zones nor by land bridges. When Wegener pushed the continents into a protocontinent, Pangaea, all the southern continents were in a south polar position suitable for glaciation. All the forests that later became the coal-producing regions of Europe and North America were contiguous with an equatorial forest zone. The book on paleoclimate by Köppen and Wegener describing some of these indications of past climates became an important reference and went through several editions. Never before had anyone offered such a comprehensive interpretation of Earth's past climates.[8] Nor had anyone before Wegener presented such a thorough proposal for the mobility of continents. But the geologic establishment of the day did not embrace the new idea.

# 3 ❖

## The World Reacts to Wegener's Idea

*If an elderly but distinguished scientist says that something is possible he is almost certainly right, but if he says it is impossible he is very probably wrong.*

—ARTHUR C. CLARKE

*If weak or fallacious arguments are mixed with strong ones, it is natural for opponents to refute the former and believe that the whole position has been refuted.*

—SIR EDWARD BULLARD

No previous suggestion of continental movements had ever evoked such strong response as did Wegener's, probably because no one had ever presented such a comprehensive body of supporting and converging evidence. As Wegener's book became widely translated, the earth scientists of the world began to see the full importance of his hypothesis, and it simply could not go unexamined or unchallenged. Other drift ideas, including Taylor's, provided little or no evidence and consisted primarily of suppositions. Accordingly, Taylor's proposal aroused little comment from the earth science community.

The fixists opposed Wegener's hypothesis primarily because it required the radical notion of horizontal movement of continents. Vertical movements were accepted because there were many known instances of land surfaces subsiding and rising again in shallow seas through geologic time. Also, land surfaces rising due to slow rebound after the removal of the weight of glaciers was a well-accepted phenomenon. Because horizontal movements were imperceptible, they seemed to require a bigger leap of faith than vertical movements that left visible evidence, such as raised and tilted relict shorelines on lakes.

Wegener's oral presentation of the continental drift concept in 1912 before

the Geological Association in Frankfurt am Main was his first appearance before a geological group. This presentation provoked no immediate reaction
because the chairman of the meeting precluded discussion and questions due
to the late hour. Apparently, Wegener's presentation went on quite a while.

In 1915, when Wegener published the first edition of his book in German,
European geologists had strong negative reactions. As translations became
available in 1924, the international community of earth scientists began to join
the chorus of criticism. Wegener's hypothesis generated negative comment
and discussion, but geologists were not inspired to take up the idea and test
the concepts independently. Criticism of new ideas is often followed quickly
with independent research by other scientists. This important element of independent testing as part of a scientific dialogue might have either advanced
the drift hypothesis or put it to rest. But significant follow-up research by others did not happen.

Wegener's optimism convinced him that all belief in fixed continents
would be abandoned within ten years after his book was published. Some
geologists expressed fascination with the innovation of the drift idea but
could not accept much of Wegener's evidence. Others rejected the hypothesis
outright and thought it unworthy of serious consideration. Many scientists
were content with the answer offered by the land bridge concept. A few were
derisive and even insulting to Wegener personally. Much of the rejection was
prompted by Wegener's lack of credentials; he had neither a degree in geology
nor geologic field experience. Hence, many felt he had no business dabbling
in geologic matters. Yet the earth science community could not leave the hypothesis alone. Each new edition of Wegener's book found additional readers
and renewed interest. Also, each new edition had additional evidence and supporting arguments, with rebuttals to some of his critics. The high level of interest, although largely negative, may have reflected an unrecognized need
among earth scientists for a new paradigm of Earth's natural history—one
without global shrinkage or land bridges.

Wegener's hypothesis was too comprehensive and covered too many lines
of evidence for any one person to disprove it totally. So critics put attention
on the details, not the overriding question of lateral mobility of continents.
The details of the drift hypothesis had numerous flaws that became easy tar-
gets for criticism without consideration of the broader issue. Nevertheless,
Wegener's hypothesis attracted enough attention among professionals that in-

ternational conferences convened for the sole purpose of discussing his concept of continental drift.

## London, 1923

One of these conferences met in 1923 under the direction of the Royal Geographical Society.[1] Attendance was by invitation and included prominent earth scientists of the time, but Alfred Wegener was not present. Statements made by presenters at the conference demonstrate the main reactions to Wegener's hypothesis. The British geologist Philip Lake took the lead in the attack. Lake accused Wegener of conveniently ignoring any facts that did not support the drift hypothesis. Speaking of Wegener, Lake said, "he is not seeking truth; he is advocating a cause, and is blind to every fact and argument that tells against it."[2] This set the tone for the entire conference. Resistance was vehement.

The strongest criticism focused on Wegener's proposal of tides and centrifugal force as the mechanisms for moving continents through the oceanic crust layer. Lake spoke for many when he said that gravity can explain movement of continents vertically due to density differences (isostatic adjustments), but "there is no force of comparable magnitude to move them sideways."[3] In his presentation Lake tore down every line of evidence Wegener had proposed.

The fitting of continents became the butt of jokes in Lake's presentation because Wegener had stretched and widened continents to smooth out folded mountains that had formed since the beginning of drifting. Wegener had justified flattening the mountains in an attempt to restore the continents to the size and shape they would have had before breaking apart in the early Mesozoic.

The matching geologic features that Wegener showed after fitting the continents back to their original positions were also questioned by Lake. He pointed out, for example, that one of the rock formations in Scotland that Wegener had mentioned, trended west-northwest to east-southeast rather than southwest to northeast as shown by Wegener. Lake also showed that Wegener had bent and rotated North America enough to make the folded Appalachian Mountains align with the folds of mountains in the British Isles. According to Lake, the folded mountains of South America and Africa were not mapped sufficiently at the time for Wegener to make a valid claim that they matched. He conceded, however, that assembling the continents into Pangaea helped resolve the difficult questions of Paleozoic fossil and glaciation distribution.

Lake's final criticism centered on Wegener's assumption that the opening of the North Atlantic could have been as late as post-Pleistocene, at the end of the ice age about 11,000 years ago. This point caused a problem for many scientists because that would be insufficient time for the northern continents to separate so widely. For some this became the basis for rejecting the entire hypothesis.

Discussion following Lake's presentation revealed some openness to the idea but with reservations. For example, from G. W. Lamplugh: "It may seem surprising that we should seriously discuss a theory which is so vulnerable in almost every statement as this of Wegener's. . . . But the underlying idea that the continents may not be fixed has in its favor certain facts which give every geologist a predilection towards it in spite of Wegener's failure to prove it. We are discussing his hypothesis because we would like him to be right, and yet I am afraid we have to conclude that in essential points he is wrong. But the underlying idea may yet bear better fruit."[4]

Another example of openness came from R. D. Oldham who had first used seismic evidence to study the earth's core. He expressed surprise that so many people regarded Wegener's theory as a novel idea. He thought there was much evidence suggesting that the continents had not always been in their present locations. But it was not yet safe for anyone to advocate such an idea. "We had the idea that all geologic features were related to the cooling and contraction of the earth and any notion of shifting of the continents was incompatible with that theory. Those ideas held the ground so strongly that it was more than any man who valued his reputation for scientific sanity ought to venture to advocate anything like this theory that Wegener has now been able to put forward."[5]

This sentiment would explain the absence of other researchers pursuing the idea of continental drift. Oldham said that the important question is not whether Wegener is right or wrong in his details but whether the doctrine of permanence of continents and oceans is right or wrong. This attitude was typical of a few who advocated a look at the big picture first—movement of continents—followed later by consideration of the details.

A Mr. Debenham agreed that Wegener's matching of geologic formations and his geodetic measurements of motion in Greenland were weak or totally wrong, but he expressed some appreciation for such a daring idea as continental movement: "Now, not for the first time perhaps, but for the first time boldly, Wegener has come forward with a theory which deals with the distribution of the continents in a bold way and offers himself for sacrifice; and he is certainly getting it. So in addition to thanking Mr. Lake for his clear undermining of the theory, I think we certainly ought to thank Professor Wegener for offering himself for the explosion."[6]

Mr. Evans agreed that the geologic map of South America was imperfect, because Evans himself had made it and knew its shortcomings. But he believed that solid geological evidence showed that Africa and South America had drifted apart. He also took the view that Wegener's hypothesis should be considered on its general merits rather than looking only at the details. He felt that the hypothesis was proposed to explain known facts, and that they must be willing to compare Wegener's hypothesis with others that were intended to explain the same facts.

The summary by the president of the Royal Society was merely a tactful rejection of continental drift, "The impression left on my mind by the discussion is that geologists, as a whole, regret profoundly that Professor Wegener's hypothesis cannot be proved to be correct."[7] He expressed a sense that a theory of continental movement might be needed, but Wegener's was not the right one. Many geologists had become uncomfortable with the cooling and contraction theory of Earth's history, and some were ready to consider mobility, but not Wegener's hypothesis.

## New York City, 1926

In November 1926 another conference convened for the purpose of discussing continental drift. This group, sponsored by the American Association of Petroleum Geologists, met in New York City. The organizer for the meeting was a Dutch petroleum geologist, W. A. J. M. van Waterschoot van der Gracht, who—apparently partial to long names—gave the meeting a lengthy title, *Theory of Continental Drift: A Symposium on the Origin and Movement of Landmasses Both Intercontinental and Intracontinental as Proposed by Alfred Wegener.*[8]

This meeting drew scientists from several countries but was the first predominantly American meeting, with nine Americans and three Europeans.

One of the Americans was Frank Taylor, who had suggested his own version of continental lateral movements, although his ideas were not under consideration at this meeting. The conference included some supporters of Wegener, as well as prominent geologists who strongly opposed the theory of continental mobility. Van der Gracht himself supported Wegener's ideas and provided supportive introductory and summation addresses that covered almost half the pages of the written report of the conference. Even though supportive, van der Gracht recognized that many questions were unanswered, especially the question of forces great enough to move continents.

Van der Gracht pointed out some of the problems with the current theory of Earth contraction and suggested that it was time for geologists to be open to alternative ideas. He pointed out that in contraction theory the Alps alone would account for enough shrinkage to shorten the earth's diameter 380 kilometers (236 miles) during the Tertiary period alone. By adding the effect of Tertiary mountain building in North America, South America, and Asia, the result would be a tremendous and sudden shrinking of the earth's crust over the past 65 million years.

A further concern about contraction theory was the lack of an explanation for large areas over the earth that seem to be unaffected by shrinking—where no mountain ranges existed. If shrinkage were uniform, the distribution of mountains should be more uniform—not so much on continental margins. In regard to alternative theories, van der Gracht felt that Frank Taylor should be given some recognition for proposing a drift hypothesis two years before Wegener—but acknowledged that Wegener deserved priority for carrying his lines of evidence much farther.

Wegener did not attend the 1926 conference but sent a short paper that contained supplementary information. Presumably, his contribution was invited as the conference was by invitation only. The presence of his paper in the published proceedings of the conference has led to a misconception that Wegener attended. Had he been there, he would have suffered through hours of criticism, much of it personal. His paper touched on the uncertainty of the glacial origins of some deposits in the tropics and predicted that the so-called glacial tills would prove not to be glacial features. Later research proved him to be right. He stated that the advantage drift theory has over other theories "is its susceptibility to verification by astronomical observations."[9]

Van der Gracht's supportive introductory paper commented on each of the main points of Wegener's hypothesis. He acknowledged that Western Australia, India, Madagascar, and East Africa had the same rocks and the same coal-bearing formations containing the fern *Glossopteris*. He also cited geological continuity on both sides of the Atlantic in at least five other good sites.

Van der Gracht held Wegener's climatological arguments as the best solution to the question of past climates and floral and faunal fossils. He saw drift as far preferable to the land bridges theory. He concluded, "I am convinced that there is continental drift, not necessarily Taylor's or Wegener's, but still intercontinental drift. If the opponents of this view, of whom I see many in the audience, and from whom we will presently hear, will openly consent to see it in this light, we may hope to make better progress in gradually approaching the truth."[10]

The open-mindedness van der Gracht asked for was not forthcoming. The supportive statements from van der Gracht carried little weight with others in the conference, and the idea of continental drift was treated as a fantasy. Established professionals have been known to refute a proposed new idea emotionally rather than reasonably, as some did in this 1926 conference. They attacked Wegener for defective thinking, writing, and research. They accused him of attacking the very foundations of their science when he had no credentials and no credibility. Of course, on this point they were right; he had no credentials in geology. He was an outsider, and this gave his opponents fuel for their rejection of anything he proposed. Being a meteorologist, rather than a geologist, apparently denied him the right to discuss geologic questions.

The greater degree of negativity at this conference relative to the Royal Society meeting of 1923 is best explained by the greater importance of fixism and crustal stability among American geologists. The Americans regarded Wegener's hypothesis as unacceptable in the face of their belief in the doctrine of uniformitarianism. The slogan of the uniformitarianist was, the present is the key to the past. If a process cannot be observed operating today, then it cannot be assumed to have operated in the past. In the light of uniformitarianism the sudden rifting of continents in the Mesozoic would be unacceptable catastrophism. To fit uniformitarianism, continental drift could not suddenly begin in the Mesozoic without any previous occurrences. (More recent studies show that continents have assembled, split and moved repeatedly throughout geologic time.) One of the leading American geologists at the meeting, R. T. Chamberlin, of the University of Chicago, spoke most strongly against Wegener's hypothesis. He questioned whether geology could truly be considered a science when it was possible for such a theory to be discussed.

Charles Schuchert (coauthor of the textbook mentioned in chapter 1) said he had tried a reconstruction of Pangaea himself and had very poor results. Thus, he could claim that his own investigations gave a fatal blow to Wegener's hypothesis. Schuchert charged Wegener with making facts fit his hypothesis by overgeneralizing and overlooking details. Schuchert's position was decidedly that of a stabilist when he said, "The battle over the theory of the permanency of the earth's greater features introduced by James Dana has been fought and won in America long ago."[11] Then he warned of some European geologists who did not hesitate to push the earth's continents and poles anywhere in order to explain certain fossil peculiarities.

Anyone who supported Wegener's ideas also came under attack from the establishment. Chester Longwell of Yale University spoke derisively of "charitable commentators" (probably referring to van der Gracht's opening remarks) who felt that Wegener's mistakes and omissions could be overlooked because he was not a geologist. Longwell concluded that some geologists might consider a different theory of Earth's history, but that evidence supporting a new theory must be much stronger than Wegener's.

> The mere fact that a group of American geologists has undertaken a serious discussion of possible continental drifting indicates a change of viewpoint within comparatively few years.
>
> Perhaps the very completeness of this iconoclasm, this rebellion against the established order, has served to gain for the new hypothesis a place in the sun. Its daring and spectacular character appeals to the imagination both of the layman and of the scientist. But an idea that concerns so closely the most fundamental principles of our science must have a sounder basis than imaginative appeal.[12]

Professor Rollin T. Chamberlin of the University of Chicago opened with a comment of faint praise for Wegener's hypothesis but raised doubts about anyone who accepted it. He suggested that Wegener's theory might appeal to some because of its simplistic conceptions. This sentiment may have been intended to intimidate any would-be supporters.

Chamberlin raised the following issues: recurring periods of mountain building were not explained by Wegener; Wegener's attempt to match glacial moraines was ludicrous; the ocean floor should have compressed into folds ahead of the moving continents; given a landmass like Pangaea, the Paleozoic floral and faunal fossils should be even more similar. Chamberlin closed with the charge that "Wegener's hypothesis, in general, is of the foot-loose type in that it takes considerable liberty with our globe, and is less bound by restrictions or tied down by awkward, ugly facts than most of its rival theories."[13]

It is useful to note that the alternative theories of that time—mountains formed by the earth's cooling and land bridges collapsing into the ocean—had almost nothing in the way of supporting facts and raised many unanswered questions. The concept of fixism was an unproven theory, and all its supporting concepts of land bridges were unprovable. Therefore, continental drift had at least equal validity as a concept for explaining known facts. Both ideas were interpretations of some basic facts, and the facts themselves were not in dispute.

Following the lead of prominent geologists of the day, other geologists also felt comfortable criticizing the drift hypothesis, most without really knowing much about it. Wegener's ideas became a subject of jokes in university classes. One favorite was about a fossil of which one half was found in North America and the other half in Europe. The two pieces fit together so perfectly, they had to have been the same animal. The implication of such stories was that Wegener was a crackpot scientist.

Some supporters argued that accepting the possibility of continental lateral movements without identifying a satisfactory mechanism is no more difficult than using a compass without knowing what causes the earth's magnetic field. A few suggested that currents in the mantle could be the cause of continental movement. Critics argued that the suggestion that currents in the mantle might be the moving force went against accepted geophysical theory of a rigid mantle, which was based on seismic observations. They felt that existing theory of the earth's interior fit known data well enough that there was no need to find another theory. Nevertheless, some geophysicists persisted in supporting the possibility of mantle currents. By 1960 an accumulation of seafloor data forced a reevaluation of concepts about the behavior of the mantle.

In 1937 the South African geologist Alexander du Toit wrote a book, *Our Wandering Continents*, which he dedicated to the memory of Alfred Wegener for his distinguished service through an innovative geologic interpretation of the earth. Du Toit discussed Wegener's ideas and added his own by suggesting two supercontinents, Laurasia in the Northern Hemisphere and Gondwana-land in the Southern Hemisphere, separated by the Tethys Sea. He offered two reasons why Wegener's hypothesis had been rejected. The main reason was the lack of a satisfactory force for moving continents. The other reason, which he deemed to be more important, was the deep conservatism of geologists of the time. Du Toit warned geologists not to hide behind conservatism without questioning the old ideas they were protecting. He urged a new look at past observations based on new theories.

Conservatism among the elder establishment is a standard element of scientific revolutions according to Thomas Kuhn's *Structure of Scientific Revolutions* (1996). Younger members of the field may be inclined to pick up new ideas but do so at some risk to their careers. Because Wegener's ideas were revolutionary, irrefutable evidence would have been necessary to get broadly based support from the more conservative geologists. Such strong evidence did not exist in the first half of the twentieth century.

At the end of the 1926 conference, van der Gracht summarized the mostly critical discussion against Wegener's hypothesis. The criticisms fell into three categories: geophysical, geological, and paleontological—in a word, everything. Van der Gracht pointed out that the majority who attacked Wegener's theory said they were not fundamentally opposed to the concept of drift. Professor Schuchert had said that geophysicists would eventually find a way to explain subsidence of land bridges. Van der Gracht wondered if they might instead find a way to explain the movement of drifting continents? He concluded that "this entire symposium stands as a token of the insufficiency of what we have been teaching in the past to explain the facts, and also of our inability to make our new attempts at an explanation fit in with everything which we now believe to be a fact either in geophysics, geology or biology. Our past confidence was founded chiefly on ignorance."[14]

Despite van der Gracht's detailed defense of Wegener's hypothesis, the earth science community remained unconvinced. Criticism by leading geologists was sufficient to sway the opinion of anyone who might be undecided. Few were willing to risk their reputations supporting a hypothesis that had become a subject of derision. After the 1926 conference, the drift hypothesis was not seriously discussed again until conferences in 1963 and 1966 when new research and new instruments led to a transformation of continental drift into a new concept: plate tectonics.

Some support of Wegener's hypothesis came after the 1926 meeting from Arthur Holmes of the University of Edinburgh. Holmes was not at the conference, but in 1928 he published a paper in *Nature* summarizing the meeting. He correctly noted that the discussions were rough and often tangential to the heart of the controversy. He agreed that much of Wegener's evidence was easy to refute, but noted that the core idea of drifting continents had not been properly discussed. The goal of seeking the truth had given way to winning a debate by crushing every argument.[15]

Holmes was a pioneer in the use of radiometric dating for finding the age of rocks and was respected for his extensive experience and knowledge. In his opinion the balance of evidence favored drift. He noted that the strong criticism during the conference was directed mainly at Wegener, and that the case for drift was actually stronger than either Taylor or Wegener had shown. In 1931 Holmes wrote an article, *Radioactivity and Earth Movements*, describing how convection currents in the mantle could cause rifting of the earth's crust with oceans forming in the newly opened spaces.[16] Perhaps if Wegener had lived in a time more conducive to interdisciplinary collaboration he would have discovered these pockets of support along with ideas for convection in the mantle as a force for moving continents. Holmes wrote that he continued to be biased against continental drift but regarded it as a fantastic idea. He suggested, however, that it would be better to call the idea an inference rather than a hypothesis.

Wegener had proposed an idea that ran exactly counter to the thought of most geologists and geophysicists of his time. For him to even suggest that the continents could move laterally was absurd in most minds. His hypothesis suggested that the earth was not rigid and that tremendous unknown forces (much greater than he proposed) were at work moving continents. He was a heretic storming the established thinking. He was surely a revolutionary, and the stir he created in earth science was nothing less than a scientific revolution.

Alfred Wegener's devotion to his drift hypothesis never wavered in spite of heavy criticism. But the next big event to capture his interest was the possibility of an expedition in Greenland. This expedition was to be the realization of an ambition and dream for many years, and it would require his full attention.

# 4 ❖

## *Preparing for Greenland*

*Let us then be up and doing, with a heart for any fate.*

—HENRY WADSWORTH LONGFELLOW

Greenland first attracted Europeans in 982 A.D. when Eric the Red found it thirty years after settling Iceland. Only 185 miles from Iceland, in clear weather Greenland is visible on the horizon by sailing only a short distance west from Iceland. Vikings found a coastline similar to Norway's with large inlets and fjords but without forests. Soon settlers began to establish permanent residences. A milder climate existed at that time, and they were able to grow some food to supplement a diet of fish.

Norse settlement probably continued in Greenland in locations on the south and west coasts until sometime in the fourteenth century. Inuit legends still tell of early settlers. Wars and plague in Europe may have stopped contact from the homelands, but little is known about the fate of the Norse inhabitants. The beginning of the Little Ice Age in the fourteenth century was probably the most significant factor in the disappearance of the Norse in Greenland. Winters became harsher, but more important, the summers became much cooler. Crops would have failed, and ice packs would have remained frozen. After several years of such hardship they would have had to abandon the place or starve.

A Norwegian document written in 1240 A.D. offers some insight into the

European perception of Greenland and explains why its climate held such interest for Europeans even at that time. "When storms come, they are more violent than in most other regions, both in coldness of the wind and the great masses of ice and snow. But these periods of bad weather usually last only a short time and come at long intervals. In the meantime the weather is fine, though very cold. For it is the nature of the inland ice to produce a continuous, cold current of air which drives away the storm clouds from its face. But the neighboring countries often suffer for this. For all the regions that are near get bad weather from this ice, because all the storms, which the glacier drives away from itself, fall upon other countries with violent gusts."[1] Present-day meteorologists recognize the additional influence on European weather from a persistent low-pressure system in the North Atlantic called the Icelandic Low.

Though sailors on Frobisher's voyage reported seeing Greenland in 1576, colonization did not resume until 1721 when Denmark sponsored settlements established by missionaries. Most of these settlements are still active today. In 1776 Denmark declared Greenland closed to outside traders other than those controlled by the government. Denmark has maintained its influence over Greenland until the present day. During World War II the United States had an agreement with Denmark to protect Greenland from foreign powers during the time that Germany occupied Denmark.

The legend of an ice-free area in the center of Greenland had persisted since the earliest coastal explorers arrived. There was no particular reason for thinking that ice would be confined to the edges, and explorers who ventured a short way inland often mentioned that they found nothing but ice all the way. But until someone actually crossed the ice, the thought of an open interior area could not be put to rest.

Inland exploration first began in the nineteenth century. One expedition in 1860 traveled a short distance onto the ice cap to investigate the possibility of laying a telegraph cable across Greenland en route to Europe. Adm. Robert Peary joined the exploration of inland ice with trips in 1883 and 1892, and in 1894 he crossed a segment of Greenland in the far north. The first to cross a wider portion of the island was a Norwegian named Fridtjof Nansen in 1882. He went across the southern end for a distance of about 500 kilometers, just over 300 miles. Nansen used five sleds pulled by men on skis traveling from east to west (another first) from August 31 to September 24. When they arrived on the west coast, Nansen built a boat from the skis, willow branches, and tent pieces, which he used to reach a settlement. There he obtained a bigger boat to retrieve the other men. Travel on the ice sheet that late in the season can be especially hazardous, as Wegener was to discover almost fifty years later.

Early expeditions had made mapping surveys of the coast and short distances inland. One of these, the 1906–1908 *Danmark* expedition, led by Mylius-Erichsen, included Alfred Wegener and J. P. Koch. Their objective was to link various surveys on the northeast coast with the American survey made by Peary in the north and northwest. Here Wegener gained valuable experience on expedition planning and survival in the Arctic. They let their ship become icebound in a fjord on the northeast coast and used it as a winter station for studies of flora and fauna and archeology. In spring 1907 two sleds traveled north along the east coast to Independence Fjord. There they split into two parties. Koch and Wegener went along the north coast across Peary Land and united the mapping done by the American expedition with those done by the Danish, German, and Austrian surveys on the east coast. Mylius-Erichsen mapped the fjord system along the east coast south from Independence Fjord. Koch and Wegener finished their work and made it back to the ship. Erichsen became stranded in a fjord by a sudden summer thaw of sea ice, making travel impossible until the fjord refroze months later. When refreezing occurred they attempted the journey, but worn-out equipment and polar nights hampered their travel. Food supplies were depleted. They became exhausted from their ascent up onto the ice cap, and Erichsen and Høeg-Hagen died. Jørgen Brønlund, a Greenlander, dragged on and found a food cache, but frostbitten feet prevented him from going further. His body was found by Koch the following spring. The other bodies were never found.

Wegener's second expedition to Greenland in 1912–1913, again with the Danes, provided additional experience on travel over the ice cap and a refresher course in survival skills. This expedition had the distinction of being the first party to spend the winter in the interior. All other winter expeditions had stayed near the coast.

Johan Peter Koch and Wegener began the trek across the ice sheet with sixteen Icelandic ponies. They planned to winter on the ice cap near the east coast, then travel to the west coast the following spring. They built a road up the front of the glacier with bridges over five crevasses. The glacier destroyed their entire effort when it calved, forcing them to start over again.[2] Eventually they reached the top of the ice. There they built a winter station called the Borg (castle) made of plywood. The structure measured 6.6 meters by 5 meters (about 21 feet by 16 feet) with prefab walls forming a triple layer of plywood with cavities between layers for insulation and with snow piled up around the sides as a windbreak. The horses lived in stables inside the building. Many crevasses crossed the path into the station. On one trip Koch fell through a snow bridge and landed forty feet below on a ledge. They hoisted him up with ropes and a rope ladder, and found that he had broken his right leg. Also, the only theodolite (surveying instrument) was lost in the crevasse. Figure 4.1 shows Wegener sitting at his desk in the winter hut.

They spent the winter collecting data—drilling and analyzing ice cores from a cellar under their hut. In the spring when the sun returned, Koch and Wegener started off across the ice with five horses and five sleds. In the begin-

FIGURE 4.1

Wegener in the living quarters at the winter station, Borg, during the 1912–1913 Danish expedition. (Used with permission of the Alfred Wegener Institute for Polar and Marine Research, Bremerhaven, Germany.)

ning the snow had a hard crust and supported the horses' weight. Every night they dug a shelter in the ice for the horses. They faced a blizzard for days. In the first forty days they had only two days of good weather, and for twelve days the weather was so bad they could not travel. The horses became snow blind, and three of them had to be killed. When they reached the interior the snow was too soft to support the horses. They sank in with each step, even though they wore specially fitted snowshoes on their hooves. Over the final stretch of the route the one remaining horse was so exhausted it was riding in the sled and the four men were pulling it. When they reached the crevasses at the west side, the horse was too exhausted to walk and they had to shoot it, just six miles from their destination.

At the west coast they still had not reached a settlement and were forced to devise a boat out of a sled covered with sleeping bag covers in order to cross a fjord. When they crossed the mountain on the other side of the fjord, they discovered the map was wrong. There was no village. Now they had to climb up to the top of the ice again. During this ascent they were hit by a storm and had to wait for thirty-five hours for a break in the weather. They were nearly exhausted and had no more provisions. With regret they shot the only dog and cooked it. Just as they began eating, a boat appeared in the fjord, which they attracted with gun shots. It was a clergyman from a nearby trading post. He took them back to the post where they got food and eventually a ship home. Wegener concluded that horses had a place in Arctic expeditions for pulling loads from the coast to the top of the ice cap, but dogs were essential for crossing the ice.

## Germany, 1928

In the early 1920s Germans struggled through what was perhaps the most devastating period of economic inflation the world has seen. From December 1920 to September 1923 the price of a loaf of bread soared from 2.37 Marks to 1.5 million Marks. The monetary exchange rate in September 1923 was 240 million Marks to one U.S. dollar. The people were in a desperate situation. In January 1924 economic and monetary reform brought some control, and the economy, though faltering, began to improve as foreign investment in industry began to pour into the country. But even as industrial production grew by 50 percent over the next five years, unemployment lingered and life for individual citizens improved slowly.[3] In this economic environment, money for a research project was almost impossible to obtain. Wegener had no realistic expectation of conducting an expedition of his own to Greenland.

In 1928 Wegener received a letter asking him to join a group of Germans in a flight by zeppelin over the North Pole. The request is indicative of the excitement generated at the time about aviation. Wegener was not interested in this idea; he considered it a publicity stunt and thought that exploration and research on the ice cap itself would be much more useful.

When a representative from a science-funding group approached him in 1928 about another expedition to Greenland, Wegener was surprised. The group asked him to conduct a summer trip to measure ice thickness in Greenland. When the group's spokesman, Professor Wilhelm Meinardus of the University of Göttingen, said that the Emergency Aid Committee for German Science would provide the funds, Wegener seized the opportunity.[4] He had long since given up hope of returning to Greenland to continue his primary line of meteorological research. Now all he had to do was write a detailed proposal telling the committee his objectives and needs.

Wegener had enough experience in Greenland by this time to know that much more than ice thickness needed to be studied. He certainly was aware that this might be his only opportunity to research all his meteorological questions about the Greenland ice cap, so he ventured to expand the scope of the research. In his proposal to the Emergency Aid Committee for German Science he included plans to measure ice thickness at several inland locations by seismic methods, determine interior elevations of the ice surface by altimeter, and verify them by surveying techniques; make gravity measurements to determine if the Greenland landmass was rising as ice load decreased; record ice temperature changes in shafts dug in the ice near the edges of the ice cap and in the interior; and observe density and structure of the firn.[5] In addition to these measurements of glacial features, Wegener planned to make daily meteorological observations over a full year plus a summer at three stations: west coast, central, and east coast at a latitude of 71 degrees north. These meteorological data would provide, for the first time, yearlong information about the atmosphere over a large ice sheet and a cross-section of the high-pressure system that had been observed during each of the earlier summer expeditions.

His 1913 expedition with Koch had made observations in the summer only, except at the east coast where they wintered.

Weather analysis over Greenland has a practical value for Europeans. Much of the weather in northern Europe is controlled by meteorological conditions in the northern Atlantic Ocean and Greenland. In 1928 little was known of those conditions. Shipping in the North Atlantic would also benefit from improved weather forecasting. Furthermore, though Lindbergh's famous trans-Atlantic flight had taken place only one year earlier, planning was already under way for regular air transport between continents. Since the most likely route for air travel to North America would pass over Greenland, the project had some immediacy. Countries on both sides of the Atlantic Ocean were eager to establish prominence in international air travel.

A basic science question related to glaciers concerned their origins and their behavior over time. Glaciers grow and advance or melt and retreat according to the fluctuations of climate, both in the short and long term. In Wegener's time scientists were very interested in the causes of climate change. The astronomic theories of Adhémar and Croll were well known, and Köppen and Wegener had recently (1924) published their book on paleoclimates showing the Milankovitch radiation curves.[6] The glacial studies Wegener proposed would provide some basic data on the behavior of the Greenland ice cap, adding another dimension to the value of Wegener's research.

### Early Climate Change Theories

In 1842 the French mathematician Joseph Adhémar suggested that ice ages might be the result of small variations in the direction of tilt of the earth's axis as it moves around the sun—similar to the wobble of a spinning top. He calculated that a slow wobble of Earth's axis took twenty-two thousand years to complete one cycle. The variation of the axis would gradually change the latitude that receives the most direct heating. Part of the time the most direct heat from the sun would be shifted a few degrees farther south and cooling would occur at the higher latitudes, possibly leading to glaciation. This idea formalized the notion of an astronomical cause for climate change, and as later investigators made their contributions, Adhémar's precession (axis wobble) continued to have a role.

By 1864 James Croll, a Scot, expanded Adhémar's idea to include the known variations in the shape of the orbit between almost circular and elliptical. This orbital change would make a cyclic variation in the distance of the earth from the sun. Croll assumed that this cyclic variation would result in snowier and colder winters. Cyclic glaciation would mean there had been repeated ice ages, rather than only one as the Swiss glaciologist Louis Agassiz (1807–1873), one of the originators of the ice age concept, had suggested. Archibald Geikie had confirmed multiple ice ages a year earlier from his studies of glacial deposits in Scotland. He noticed that at least four glacial deposits could be seen, and the extent of weathering of the lower deposits suggested they were considerably older than the upper layers. This gave early support to the concept of climate cycles due to orbital effects.

One tantalizing detail was still unanswered. Did the glacial deposits correlate in age with the cycles of Croll and Adhémar? The time of the ice age based on orbital variations of Earth could be calculated, but geologists had no way to date the actual ages of the glacial deposits. They could only make a guess on their ages based on assumptions about the rate of weathering for the older deposits.

American geologists contributed to the age question when they estimated a rate for the retreat of Niagara Falls. Assuming that the retreat of the falls could not have begun until the ice melted away, they concluded that ice had left the area as recently as 10,000 years ago. Since Croll had estimated that the last ice age ended between 80,000 and 150,000 years ago, his theory began to lose validity as an explanation for climate change. When his theory failed to explain what was seen on the ground, it became something of a historical curiosity, but not an acceptable explanation for glaciation.

Milutin Milankovitch (1879–1958), a professor of mathematics, physics, and astronomy in Belgrade, began after 1909 to pursue a work of great magnitude and major importance to science. He decided to work out formulas for calculating the temperatures at any latitude based solely on changes in the shape of Earth's orbit and variations in the orientation of the axis. He extended the work of Adhémar and Croll to include changes in the orbit as well as axis wobble. By 1924 Milankovitch had completed enough calculations to revive the astronomic approach to climate change theory.

Croll had advanced the possibility of climatic effects due to orbital changes, but he lacked the mathematical skill to deal with the problem accurately. Other mathematicians had calculated how the orbit shape would vary over time, but none had approached the effects of these changes on heating of the earth. Isaac Newton had shown that Earth's heating depends on its distance from the sun and the angle of the sun's rays hitting the surface. Milankovitch worked on the problem of radiation and heating effects for about thirty years, without computers or electronic calculators.

His colleagues thought it strange to be calculating temperatures by such methods when there already existed a world network of weather observation stations providing direct readings of daily temperatures. The applications for such calculations soon became obvious. If the expected present-day temperature at each latitude on Earth could be calculated, then it would be possible to calculate the temperatures for each latitude at any time in Earth's history—a valuable boost to paleoclimatology. Furthermore, temperatures could be estimated for planets and moons in their various orbital and axial positions. This freed climatologists from the limitations of direct observation and allowed them to consider past Earth climates as well as planetary climates.

Milankovitch's first results were published in 1920 when he was forty-two years old. In this publication he gave a mathematical account of the present climates of Earth, Mars, and Venus. He also demonstrated that astronomical variations on Earth were sufficient to produce ice ages just by changing the distribution and timing of the amount of sunlight on earth. He found that the amount of radiation received at high latitudes had greater variation and influenced climate more than in the tropics. He showed that the eccentricity of Earth's orbit changed from nearly circular to more elliptical on a 100,000-year cycle and that this determined the effect of precession on radiation received on Earth.

He showed that precession of the axis followed a cycle of 19,000 to 23,000 years, with an average of 22,000 years to complete one wobble. Other scientists have since calculated the precession cycle at 26,000 years. Today the North Pole points to the star Polaris. Four thousand years ago it pointed midway between the Big Dipper and the Little Dipper. Six thousand years ago the North Pole pointed to the tip of the handle of the Big Dipper.

Milankovitch also considered the effect of a change in tilt of the axis by about 1.5 degrees on either side of its present 23.5 degrees on a cycle of 41,000 years. All these variables were figured into his calculations of the changes over time of radiation at any latitude on earth. He considered that the total radiation received by Earth and other planets from the sun had not changed significantly with time.

Milankovitch's book did not receive much notice when it came out in 1920. Perhaps the title, *Mathematical Theory of Heat Phenomena Produced by Solar Radiation,* was a bit too heavy for many readers. However, it immediately attracted the attention of Wladimir Köppen, who by this time was working on his book with Wegener on Earth's climates of the geologic past.[7] Köppen soon contacted Milankovitch, and they began a long-lasting correspondence and collaboration. Milankovitch reported his excitement at having affirmation from such a well-known scientist as Köppen. In their correspondence Milankovitch related that his next task was to calculate Earth's past temperatures. This prompted Köppen to invite Milankovitch to collaborate on his forthcoming paleoclimate book by developing summer radiation curves. He also invited Milankovitch to come for a visit to discuss the matter.

Croll and Adhémar both had assumed that winter temperatures at high latitudes controlled the cooling leading to an ice age. Milankovitch wanted to consider the validity of all assumptions, so he asked Köppen for his opinion on the significance of seasonal temperatures. Together they discussed this issue at length, and Köppen ultimately concluded that summer temperatures are more significant than winter in creating a glacial climate. Köppen reasoned that even winters in our present climate had temperatures cold enough for snow to accumulate, but that it all melts in the summer, except for a few locations already having glaciers. If solar radiation decreased during the summer enough to inhibit melting, then snow would begin to accumulate, eventually producing glaciers.

Milankovitch used this assumption in his calculations of summer radiation at the northern latitudes of 55, 60, and 65 degrees over a period of 650,000 years. His summer curves showed low points of radiation that he identified as four possible ice advances in that period. Köppen wrote back that Milankovitch's radiation lows agreed amazingly well with the history of alpine glaciers devised by the German geographers Penck and Brückner. They had estimated the number of glaciations and the time intervals between glaciations by counting the number of glacial deposits and the amount of weathering of each one. Their results showed four pulses of ice advances with long intervening warm periods.

This correlation of theoretical work with fieldwork provided the most valuable aspect of Milankovitch's radiation curves. Now field geologists had a prediction, based on theory, of the number of layers of glacial deposits they should expect to find. The radiation curves also predicted the age for each deposit. This provided a calendar, with dates, for the Pleistocene ice events in Europe and North America. In 1924 Milankovitch's curves appeared in Köppen and Wegener's book *Die Klimate der geologischen Vorzeit (Climates of the Geologic Past)* and received widespread notice in the scientific world. Though Milankovitch's radiation curves stimulated discussion and field research, some skeptics doubted that winters in a glacial period would actually be milder than winters today. Analysis of ice cores taken in Greenland in recent years supports Milankovitch's conclusion that milder winters occurred during glacial periods.

In the late 1930s radiocarbon dating techniques provided a strong tool for dating vegetative remains found in glacial deposits and associated bogs. Results from this new technique indicated the most recent glacial maximum to be 18,000 years ago, rather than 25,000 years as Milankovitch predicted. Milankovitch had suggested that a lag time between a radiation minimum and a glacial maximum would occur. Indeed, lag times between radiation peaks or lows and a corresponding temperature response are common. Nevertheless, the discrepancy in radiocarbon dating of the most recent glaciation and Milankovitch's radiation curves weakened support for his method. Because radiocarbon dating has a limit of about 40,000 years, actual dates of earlier glaciations could not be determined.

Later dating methods using radioactive isotopes of uranium, thorium, and potassium extended the dating capability further back in time. In 1956 a new method using thorium was used to date coral reefs as old as 150,000 years. This had climatic implications because corals require a narrow range of water temperature and depth. Therefore, past changes in sea level and water temperature could be detected. As improving technology allowed older events to be dated, Milankovitch's curves began to pass more tests based on field data.

## Final Preparations for the Greenland Expedition

Wegener developed a grand proposal covering a preliminary reconnaissance trip in 1929, and sixteen months of meteorological, geodetic, and glacial data gathering in 1930–1931 for a sum of 500,000 Marks (U.S. $120,000 at the November 1929 exchange rate). The Emergency Aid Committee for German Science quickly approved the portion of Wegener's proposal that provided for a preliminary reconnaissance in 1929, one year ahead of the main expedition. In the preliminary expedition, Wegener would select suitable locations for the west and east stations and find a passable route for ascent of the glacier onto the ice cap surface. He would also make arrangements with Greenlanders for dogs, dog food, and laborers when the expedition arrived the following year. In the reconnaissance trip there would be a chance to make some preliminary scientific observations and test some new instruments.

At the end of March 1929, Wegener departed with Johannes Georgi (who had also been on the 1912 expedition with Koch and Wegener), Ernst Sorge and Fritz Loewe for Greenland on the *Gertrud Rask* (Figure 4.2). A suitable site to ascend the glacier had to meet several criteria. First, the glacier must have a slow rate of movement so that calving would not be a problem. The

FIGURE 4.2
Johannes Georgi, Alfred Wegener, Fritz Loewe, and Ernst Sorge
aboard ship. They are returning from the 1929 reconnaissance trip
made in preparation for the 1930–1931 expedition to Greenland.
(Used with permission of the Alfred Wegener Institute for Polar
and Marine Research, Bremerhaven, Germany.)

expedition with Koch in 1912 was nearly wiped out by glacial calving. Another condition was that the front edge of the glacier must be near the shore so it would not be far to move sleds and equipment over bare rocks in the summer. Also the glacier must be neither too steep for an ascent nor too severely crevassed along the ascent route.

The search soon focused on Umanak Fjord (Uummannaq) on the west coast, and only one of the glaciers feeding into Umanak Fjord was suitable for ascent.[8] This glacier, the Kamarujuk (Qaumarujuk), descended from the ice cap (at an elevation of 1,000 meters (3,280 feet)) down to sea level over a distance of about 4 kilometers (2.5 miles), ending 300 meters (984 feet) from the sea in a flat outwash plain. An icefall (a sudden change in the slope of the ice surface) was present part way up the ascent route, and above the icefall the glacier was greatly crevassed. Both these features would make the ascent more difficult, but it was still better than the alternative routes. At the top of the glacier was the Scheideck Nunatak, and near its base they established a site for the West Station.[9]

On the 1929 expedition Wegener's party made several dogsled trips inland from the west on the ice cap to test the consistency of the ice surface. They also made the historical first measurement of ice thickness using the seismic method. Their measurement of 1,113 meters (3,651 feet) at 42 kilometers (26 miles) from the west coast provided the first knowledge of ice thickness on Greenland. Today the Greenland ice cap is known to be as thick as 3,350 meters (about 11,000 feet) in central areas.

A major innovation in Wegener's plan was to use propeller-driven motor sleds in addition to dogs for transporting supplies to the Central Station in the main expedition in 1930. Motor sleds were a new idea for polar transportation and had not yet been tested in an Arctic winter environment. The potential advantages were sufficient to make the motor-driven sleds an attractive option. Unlike dogs, which would have to be fed all winter, the motor sleds would need only enough fuel to get there and back; they would require no fuel when not in use. The biggest advantage of motor sleds was that more trips carrying more weight could be made to provision the Central Station. Wegener estimated that the motor sleds could make the round-trip in six days, whereas dogsleds required three or four weeks. Given the short time in summer suitable for travel, time was a critical consideration. Much of the excitement about the expedition related to testing these sleds.

The sleds were manufactured by the Finnish State Airplane Factory and originally designed for winter transport over sea ice to offshore islands and on Finnish lakes. Wegener's adviser on engineering matters, Asmus Hansen, went with Kurt Schif, who was to be in charge of assembling the motor sleds in Greenland, to visit the factory in Helsingfors, Finland. They were pleased when they saw the bright red sleds finished and ready to be packed in large wooden crates. The sleds resembled a small airplane body with a rear propeller. They ran on four hickory skids, and the front skids steered like a car. The following year those sleds played a major role in difficulties faced by the expedition.

With these on-site decisions settled, Wegener's preliminary expedition ended and they returned to Germany in November 1929. Upon their return, Wegener learned that stock markets had crashed, and economic difficulties in Germany had worsened. When he returned to Berlin he was told that the expedition would be canceled or at least postponed for a year. This news was devastating. However, soon after the market crash optimism rose because of predictions that the slump would be short-lived. The optimism resulted in restoration of the expedition funds in December, and by early 1930 preparations for the full plan began in earnest.

Wegener began shopping for supplies and equipment. Purchase of the propeller-driven sleds, radios, and photographic equipment was delegated to Johannes Georgi and Ernst Sorge. The major task, however, was finding and hiring the necessary scientists and technicians who must be well qualified and physically fit. In addition, expedition personnel must have a spirit of adventure with the ability to endure heavy labor and hardship in a difficult and dangerous environment. They must be willing to endure privation, hunger, and cold. They had to share in the hard work of transporting loads over the ice, putting aside their personal scientific interests until the transport was completed. They had to understand that they might put their lives at risk just by living and working in an ice cap environment.

Despite all these constraints many men applied for the jobs. Table 4.1 is the final list of expedition members, their experience, and job assignments. As these men began working together, they soon developed a high respect for Wegener. Johannes Georgi wrote that Wegener's trust in the men and his congenial manner soon created a strong comradeship among the men.

Else Wegener wrote of her reluctance to see Alfred leave on a fourth ex-

TABLE 4.1 *Personnel on the 1929–1930 expedition to Greenland*

| Name | Professional position | Expedition assignment |
|---|---|---|
| Alfred Wegener | Professor of geophysics and meteorology, Graz University | Director; ice cap climate, glaciology |
| Johannes Georgi | Naval Observatory, Hamburg | Meteorologist in charge of Central Station |
| Rupert Holzapfel | Austrian meteorologist | Meteorological measurements at the West Station |
| Fritz Loewe | Aviation Weather Bureau, Berlin | Glaciological work at the West Station |
| Ernst Sorge | Secondary school teacher, Berlin | Ice thickness measurements at Central Station |
| Karl Weiken | Geodetic Institute, Potsdam | Gravity measurements and geodetic survey from the west coast to the Central Station |
| Kurt Wölken | Geophysical Institute, Göttingen | Ice thickness measurements at West Station and on the ice cap |
| Kurt Herdemerten | Engineer, Düsseldorf | Blasting and shaft digging operations |
| Hugo Jülg | Middle school teacher, Linz, Austria | Assistant surveyor |
| Georg Lissey | Engineering student, Hamburg | Assistant surveyor |
| Emil Friedrichs | Skilled mechanic, Naval Observatory, Hamburg | Engine maintenance on *Krabbe* |
| Franz Kelbl | N/A | Portable buildings assembly, motor sled driver, and radio operater |
| Manfred Kraus | N/A | Portable buildings assembly, motor sled driver, and radio operator |
| Kurt Schif | Aeronautical engineer, Aviation Research Institute, Berlin | Direct transport and assembly of motor sleds in summer of 1930, then return to Germany |
| Vigfus Sigurdsson | Icelander | Purchasing and managing Icelandic ponies |
| Jon Jonsson | Icelander | Helping with ponies |
| Gudmunder Gislason | Medical student from Iceland | Helping with ponies |
| Bernhard Brockamp | Scientist, Askania Works, Berlin | Came in summer of 1931 to make measurements of ice thickness |
| Walther Kopp | Meteorological Observatory, Lindenburg, Germany | Meteorologist and leader of East Station |
| Arnold Ernsting | Engineering Student, Darmstadt | Assistant at East Station |
| Hermann B. Peters | Zoologist, Kiel, Germany | Assistant at East Station |

*Source*: E. Wegener, ed., *Greenland Journey*, 9.

pedition, but she said nothing to dissuade him. She worried that something might go wrong with the expedition. "Would everyone support the objectives of the expedition, or would individual egos upset all the careful planning and endanger the lives of all? My heart was heavy."[10] Else wanted nothing to stand in the way of Alfred's reaching his lifelong goal.

# 5 ❖

## Arriving in Greenland

*One misfortune never comes alone.*

—HENRY FIELDING

M oving an Arctic expedition with ninety-eight tons of goods contained in twenty-five hundred crates, bags, boxes, and barrels from Copenhagen to western Greenland required an immense effort. Add to that fourteen men embarking at Copenhagen, picking up three more men and twenty-five ponies in Iceland, and transportation difficulties became enormous.

Alfred Wegener planned the entire expedition based on his experience with three earlier trips to Greenland. He delegated some purchasing tasks to Fritz Loewe and Johannes Georgi, but every detail received Wegener's attention. He worked as carefully and methodically in expedition planning as he did in his research.

Finally, all preparations completed, Wegener arrived in Copenhagen to oversee the loading of the entire expedition aboard a large Danish ship, the *Disko*, and on April 1, 1930, they left Copenhagen for Reykjavik, Iceland. There, on April 10 they took twenty-five ponies on board along with Vigfus Sigurdsson, Jon Jonsson, and Gudmunder Gislason who came to care for the ponies and to lead the pack trains up the glacier. Sigurdsson had accompanied Wegener and Johann Koch on the 1912–1913 expedition across Greenland.

On the fifteenth they reached Holsteinborg (Sisimiut) in southwestern

Greenland where the melting ice had already opened the southern coast to shipping. Here the crew had to offload everything onto the pier and await the arrival of a smaller ship, the *Gustav Holm*, which was more suited for navigation among the ice floes. The unloading went smoothly although one of the ponies slipped off the edge of the ship and fell into the icy water. The pony was injured slightly in its panic, and Vigfus jumped in to guide it to shallow water. After being thoroughly dried and warmed, man and horse were ready to continue.

By the night of the nineteenth the *Gustav Holm* had arrived at Holsteinborg, and the crews began loading as much of the expedition equipment as possible. Some of the goods had been damaged in unloading. Gasoline drums had been dented, holes punched in boxes, and bales of hay dropped into the sea by the inexperienced expedition members. In five days the freight filled every cranny of the *Gustav Holm*. Drums of kerosene shared space in the ship's coal bunker. Gasoline cans jammed the space on the main deck. A small boat sat atop gasoline cans with more gasoline cans inside the boat. Dynamite boxes mixed with the ponies' hay, and detonators filled the lifeboats. Wegener wrote: "We have a damned risky cargo on board. If fire breaks out we're done for; no hope of putting out gasoline. The only consolation is that we will have a very imposing and expensive cremation ceremony."[1]

The huge packing crates containing the motor-powered sleds just barely fit into spaces on either side of the main hatch. The ponies were also on the main deck next to the power sled crates. Crates of food were loaded in the most inaccessible place on the bottom deck under bales of hay. After everything was loaded, Loewe realized he had to disembark early on a reconnaissance dogsled journey, and he needed six boxes of food for that trip. To find them he had to burrow under the hay in the hold and pull the boxes out one by one. The map in figure 5.1 includes the Greenland locations pertinent to the expedition.

Before the ship left port, the town leaders of Holsteinborg came on board for a farewell visit. Wegener served them rum toddies while Wölken and Holzapfel entertained them with harmonica and guitar. When the ship got under way, the overcrowded conditions forced many men to sleep on the floor of the lounge or belowdecks in the hay.

When they arrived in the town of Godhavn (Qeqertarsuaq) about a hundred miles north of Holsteinborg, the governor of North Greenland told them that the sea ice pack was still solid to the north along the west coast. This meant the expedition could not reach the Kamarujuk Fjord to unload for the trek up

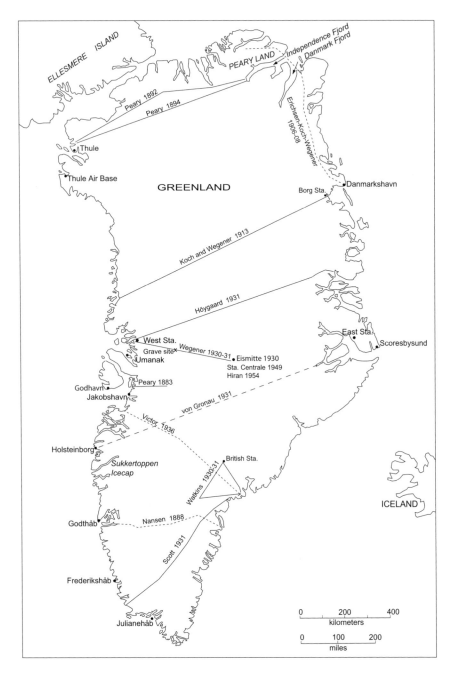

FIGURE 5.1

Locations of selected research stations and exploration routes in
Greenland. (Adapted from Fristrup, 1966)

the glacier in order to build the West Station. The temperature was -16° Celsius (3° Fahrenheit) and a heavy snow was falling. This delay, with no sign of warming, was the first disappointment for the expedition. On the last day of April, Greenland's summer seemed very far away, and members of the expedition abandoned their European clothing and donned Greenlander furs.

The *Gustav Holm* proceeded north in open water to a point at the edge of the land-locked ice, several miles west of the settlement of Umanak on Umanak Island in Umanak Fjord (Uummannaq). A dogsled from the settlement came alongside, and Wegener sent a letter to the leader of the settlement explaining who he was and the nature of their expedition. A few hours later the settlement leader arrived and came on board. Wegener learned that the ice extended north of the Kekertat Islands (Qeqertat) with no opening into the Kamarujuk Fjord.

In a small boat Wegener and Vigfus Sigurdsson went to look for a site at which to begin unloading onto the ice. They selected a spot near Kekertat Island. The *Gustav Holm* crew moved the ship to this place, anchored along the ice, and began unloading onto the ice about 16 kilometers (10 miles) out from the settlement of Uvkusigsat (Uvkusigassat) (figure 5.2).

The leader of Uvkusigsat arranged for Greenlanders from his and several other nearby villages to come with twenty dogsleds to help move the goods over the sea ice to Uvkusigsat.[2] Wegener paid the Greenlanders by the pound for goods carried, so they piled their sleds to the limit for the running trek to Uvkusigsat. This effort was risky as the ice had begun to soften, and in places it was soft enough to push a stick through. This was a special problem for the ponies, which were led safely across the ice in two groups tied in lines. The heaviest items, such as the small boat and the motor sleds, had runners attached to them so they could be pulled by two ponies.

The closer they got to the shore the softer the ice became. Still a mile from shore the Greenlanders stopped and said they would go no farther. The ice was sagging under the weight of the sleds and a small lake was forming in the sag. The sleds were in danger of sinking out of sight if they stayed in one spot. Wegener promised the Greenlanders a luxurious meal with generous servings of schnapps if they would continue carrying loads over the ice. The schnapps did it. The last part of the trip had the dogs half running and half swimming while everyone helped push from behind.

In three days the entire load sat on dry solid ground near Uvkusigsat, still

FIGURE 5.2
The *Gustav Holm* unloading the expedition equipment at the
edge of the pack ice, May 1930. The large crate held the propeller
sled parked to the right of the crate. (Used with permission of
the Alfred Wegener Institute for Polar and Marine Research,
Bremerhaven, Germany.)

26 kilometers (16 miles) from the base of the Kamarujuk glacier (figure 5.3).
On May 10 the *Gustav Holm* departed with signal flags flying a farewell message and the small deck gun firing a salute. Tides and wind had created an ice
surface that was too rough for regular transport by dogsled across the fjord, so
the expedition had to wait for the ice between Uvkusigsat and the head of the
Kamarujuk Fjord to thaw and break up. Once the fjord was open for travel,
they would carry everything in their small boat, the *Krabbe* (figure 5.4) to the
starting point and begin the ascent of the Kamarujuk glacier onto the surface
of the ice cap. They set up two tents as summer residences near Uvkusigsat
and settled in to wait for the thaw. Wegener felt satisfied with the expedition's
progress so far. "I am sitting on my sea chest in our fine big wooden floored
tent with my back against the bunk. In front of me is a table whose drawers
already hold knives, forks, spoons and cups for six occupants of the tent and
three visitors. Vigfus is cooking pemmican on our brand new Primus, and in
addition we are burning one of the numerous kerosene stoves which was presented to us. On the wire mattresses of the bunks lie sacks filled with hay, and

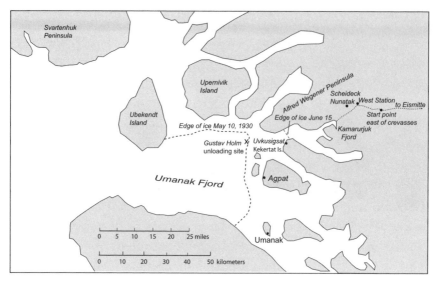

FIGURE 5.3

Umanak Fjord showing locations mentioned in text. Kamarujuk Fjord was the starting point for climbing onto the ice cap. Note the peninsula that was later named for Wegener. (Adapted from E. Wegener, ed., *Alfred Wegeners letzte Grönlandfahrt*, 1932.)

FIGURE 5.4

The *Krabbe* provided water transportation during the 1930–1931 expedition. (Used with permission of the Alfred Wegener Institute for Polar and Marine Research, Bremerhaven, Germany.)

on top of that fine new down sleeping bags. Outside hangs a thermometer that reads ½ degree below zero [31° Fahrenheit], but the tent is quite warm."³

## Waiting for Sea Ice to Thaw

Optimism and a sense of excitement filled the camp when the Greenlanders estimated that the ice would break up within eight to fourteen days depending on the daily temperatures and winds. The waiting time was put to use testing instruments and repairing a few pieces of equipment damaged during the trip. Radio operators strung an antenna 500 meters (1,640 feet) long up a cliff face and began receiving time signals by radio. The occupants of the tent took turns cooking meals that consisted mostly of oat porridge and pemmican.⁴

After ten days the weather was still stormy with heavy snowfall and mountains hidden in mist. No sign of ice breakup appeared in the fjord. Two days later the storm cleared at midday, and the wind had begun breaking the ice to the west in open water—not in the fjord. Using the open water Wegener and Lissey took their small motorboat, the *Krabbe*, around the Kekertat Islands all the way to Umanak where they talked with the village leader and visited the captain of the *Hvidfisken*, which was contracted to deliver some goods when the Kamarujuk Fjord opened. Soon after they left Umanak, an east wind jammed its harbor with floating ice making it impassable.

On the way back to Uvkusigsat in the *Krabbe*, Wegener and Lissey rescued three Greenlanders who, with their sleds and dogs, were marooned on the sea ice that had separated from the land. When they reached the land-locked ice they secured the boat and made a hasty trip through standing meltwater against a raging headwind. After half an hour they reached Uvkusigsat, splattered and wet from following the running sled dogs.

Wegener, Sorge, and Kraus went on a reconnaissance by dogsleds to the Kamarujuk Fjord and up the glacier to the Scheideck Nunatak near where West Station was to be located. This required a tedious trek over rough sea ice from Uvkusigsat to the head of the fjord, then four miles up the steep slopes of the glacier to Scheideck. The climb up the glacier was especially strenuous along the section they referred to as the "break," where the glacier forms an ice fall as it goes over the steepest part of the mountain. When they later had to move all the equipment and supplies to West Station, traversing this ice fall would become a major hurdle.

Returning to Uvkusigsat, Wegener and the expedition fell into a state of gloom. The weather was still bad much of the time, and the ice had not yet broken up. They had been waiting for the thaw for thirty-four days, and the lagging schedule concerned everyone. So much needed to be done, and valuable weeks of the short Greenland summer were slipping away. During this time, many entries in Wegener's diary began with the somber words, "Waiting day."

The *Hvidfisken* arrived from Umanak with additional supplies Wegener had purchased, and the ship tried to break through ice to reach the harbor at Uvkusigsat. Still solid, the ice could not be moved. The expedition members worked all night through the warm midnight sun of June 17—just a few days before the summer solstice. They sorely needed an east wind to make the ice move, but the sea and the wind remained calm. They tried a series of dynamite charges to break a lane through the ice to no avail. They continued from 6 P.M. until 5 A.M. blasting and ramming with the ship. Suddenly the ice began to open, but not in the place where they had worked so hard. The wind and tide had caused a break thousands of meters long, and the Kamarujuk Fjord was now open. Wegener wrote, "All our work had been wasted. We might as well have slept."[5]

After thirty-eight days of waiting, the *Krabbe* dropped anchor in Kamarujuk Fjord and prepared for the massive movement of all the expedition's goods from Uvkusigsat to the head of Kamarujuk Fjord and up the glacier to Scheideck Nunatak to establish West Station. The portion from Uvkusigsat to the foot of the Kamarujuk Glacier required only a 26-kilometer (16-mile) shuttle by motorboat. All twenty-five hundred boxes and barrels had to be moved to the head of the Kamarujuk Fjord and stacked well away from the water levels of the highest tides. When they got to the terminus of the glacier they felt great relief, but when they looked at the ice fall on the glacier ridges and crevasses above them, their spirits sagged.

The expedition members pitched their tents, and the Icelanders built wire enclosures for the ponies. At this time of year no shelter was necessary for the ponies as the sun was always shining and the temperatures were mild. The days were very hot when the sun shone directly into the fjord from the south and southwest, but became cool when the sun moved behind the three-thousand-foot high cliffs on both sides of the fjord.[6] Part of the day they worked stripped to the waist.

Between the shore at the head of the fjord and the terminus of the glacier the terrain was a flat surface about 300 meters wide (nearly 1,000 feet) of rock and gravel outwashed from the glacier and furrowed with meltwater streams. The glacier was smooth and sloped in a gentle curve up to about 350 meters (1,150 feet) above sea level. There the ice surface steepened where the glacier moved over a cliff in the underlying mountain creating an ice fall. At this point crevasses opened as the ice bent over the precipice. The ice fall was a hurdle, but not impassable. Expedition members spent days of hard labor building a path chipped from the ice with picks and shovels. Ponies would be using the path to pack crates of provisions, cans of fuel, and pieces of a portable building up to the West Station site. Where possible the ponies could also pull sleds full of goods. Three tons of provisions and equipment would then be moved on to Central Station, which Wegener named Eismitte (Mid-Ice), 400 kilometers (250 miles) eastward in the middle of Greenland. Figures 5.5, 5.6, 5.7, and 5.8 give some idea of the labors involved in this phase of the expedition.

Road building along the glacier began with ice chips flying. Narrow crevasses were filled with powdered ice; others had to be bridged with planks. Most of the work was done from 6 P.M. to 6 A.M. to avoid the heat of the sun and the glare on the ice. Above the ice fall, the glacier was still covered with snow and was too difficult for the pack ponies. There they switched to dogsleds with only half loads, as the trail was too steep for the dogs to pull a fully loaded sled. In a few days pack ponies, dogsleds, and the motorboat, *Krabbe*, formed a continuous flow of goods from Uvkusigsat across the fjord and up the glacier to the West Station site. Loewe and Holzapfel had already set up a big tent for meteorological equipment to begin observations at West Station. The return trips down the glacier provided the men some wild and reckless rides in empty sleds.

Nothing seemed to stay the same for the busy crew. The glacial ice melted during the heat of the day, changing the condition of their handmade path. The meltwater streams got deeper, the crevasses got wider, and the terminus of the glacier steadily retreated. The storage depots were in danger of being engulfed by mud. Melted and eroded ice paths had to be chopped out anew daily. In one mishap on the slippery ice, a loaded pony fell into a crevasse and

FIGURE 5.5
The motor sleds had to be winched up the front of the glacier. The engines were brought up separately. (Used with permission of the Alfred Wegener Institute for Polar and Marine Research, Bremerhaven, Germany.)

FIGURE 5.6
Transporting loads by Icelandic ponies required men to lead the ponies and guide the sled. (Used with permission of the Alfred Wegener Institute for Polar and Marine Research, Bremerhaven, Germany.)

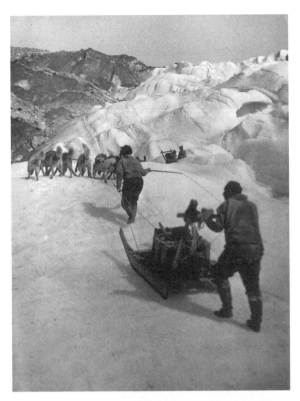

FIGURE 5.7
The small loads for each sled required many trips up the glacier front by (*a*) dogs and (*b*) human power if all the dogs were in use. (Used with permission of the Alfred Wegener Institute for Polar and Marine Research, Bremerhaven, Germany.)

FIGURE 5.8
Ernst Sorge steps across
a crevasse. These
hazards presented a
major problem for
transportation. (Used
with permission of the
Alfred Wegener Institute
for Polar and Marine
Research, Bremerhaven,
Germany.)

died. The expedition took advantage of this accident and had their first fresh meat since leaving Copenhagen.

They continued to sleep during the day but complained that the sun made the tents like hothouses, which resulted in poor quality sleep and increasing fatigue. They had a breakfast of coffee and porridge about 8 P.M., then began feeding and harnessing the ponies for a night's work. The load put on a pony's back was supposed to weigh no more than 90 kilograms (200 pounds), but to hurry the operation they had been increased by about one-third. A hazard of working with loaded ponies was that they were easily startled and would sometimes bolt off the trail. In one incident the ponies were crossing a meltwater stream bridged by planks. The hollow noise of their hooves and the rushing water below made the ponies nervous on such crossings. The last pony in the string suddenly got excited and bolted forward ramming into another pony, causing the second pony to slip. The second pony fell into the stream and became jammed between the walls of ice with his load still attached. As the fallen pony was still attached to the pony ahead, that one was also in danger of falling into the stream. Since the ponies that had already crossed the planks were standing on ice, their footing was unstable, and the whole string of five ponies could be pulled into the stream.

Holzapfel quickly saw what needed to be done. He cut the rope between the last two ponies, and jumped into the stream to release the load still bound to the fallen pony. When the load was released, the excited pony jumped up knocking Holzapfel down. Holzapfel was swept downstream some distance along the slick icy streambed. He finally stopped himself and climbed out of the stream dripping wet and bleeding from cuts on his face. The pony was also bleeding from a cut on the leg but was not badly hurt. Fortunately, all this happened before they had begun the ascent of the ice fall. Both horse and man would have been in much greater danger on those steeper slopes.

Another incident caused by a pony slipping on the ice is recounted in Georg Lissey's journal:

> The path has been cut obliquely across the steep ice slope. Here the ponies proceed cautiously, choosing their footholds carefully. On the left they occasionally scrape against the cliff; on the right there is a steep slide as smooth as a mirror, ending suddenly in a 15-meter [50-foot] drop.
>
> At this very spot a meddlesome dog pushes past the last pony, which shies, misses its footing, falls and slides headfirst down the slope. Just at the edge of the precipice its hooves get a hold on the ice. With a jerk the animal gets to its feet and stands trembling with its head hanging over the edge of the crevasse. Its hind legs are much higher than its fore-legs and the load has slipped forward almost to its neck. Gudmund immediately jumps down, sticking his crampons hard into the ice. He pats the pony on the neck and speaks soothingly to it. The animal moves in fright and pins down Gudmund's foot. If the pony slips now it will take Gudmund with it. Soon Jon is beside him. They carefully lift off the loads. This is no joke to lift boxes weighing 60 kilograms [132 pounds] off a terrified, shaking pony in such a position on a steep smooth ice slope at the extreme edge of a precipice.

The boxes are thrown into the crevasse. Jon then gets behind the
pony, grips it by the tail and pulls it back several meters. Gudmund
pulls the pony around by the halter a few steps and it is safely on the
trail again.[7]

The two propeller-driven motor sleds were far too heavy for dogsleds. The
large crates, 5 by 2 by 1.5 meters (16 by 6 by 5 feet), were unloaded so the
sleds could be pulled by ponies on flat or gently sloping ice. They had to be
pulled up the steep ice fall by hand-operated winches and cables. This was
slow and tedious work, but fortunately there were only two motor sleds. They
had to move the winches sixteen times before reaching the top edge of the ice
cap at an elevation of 900 meters (2,950 feet).

Week after week the movement of material up the glacier continued.
They worked by night and tried to sleep in the hot tents during the day. Because of the thirty-eight-day delay getting started, they always had a sense of
being behind schedule—creating a constant sense of urgency. There was still
much to do. They not only had to move everything to the top of the glacier at
Scheideck Nunatak but also had to make at least three sled journeys to Eismitte carrying provisions sufficient for three men to spend the winter. They
needed to erect the portable buildings for housing and laboratories at the West
Station where ten men would overwinter. The building and occupants had to
be supplied with bunks, tables, utensils for cooking and eating, food, kerosene
for cooking and heating, scientific instruments, explosives, and hundreds of
other items. Everything needed for the coming winter of 1930–1931, as well as
for the following summer, had to be taken to the West Station now. After
everything had been moved, no further equipment or provisions could be
transferred because they planned to slaughter all the ponies after they completed the work. Hay supplies would not last through the winter, and the
ponies' meat would provide food for men and dogs.

As the ice path continued to melt, the crew made an alternative trail along
the lateral moraine that paralleled the edge of the glacier.[8] This trail had a
steep climb to the crest of the moraine ridge, but once there, ridge travel was
easier than on the glacier. Gradually the ice path was used less, and the
moraine trail was used more. By mid-July the ice fall could no longer be
crossed because melting had made the ice too dangerous. Also, by this time
enough had been brought up to West Station that Georgi, Weiken, and Loewe

could begin packing provisions and equipment for the first transport of goods to Eismitte 400 kilometers (250 miles) away.

Despite the great quantity of goods transported, there was already a need to begin resupplying. Wegener went to Umanak on August 2 to buy more ice crampons, milk, dried fruit, flour, and two thousand cigarettes. That day he received a telegram from Walther Kopp, leader of the party sent to establish the East Station in Scoresby Sound. They were still waiting for the sea ice to break up on the east coast and had not yet reached the site where the East Station was to be established. Furthermore, the party was concerned that they might not get there in time to establish a functioning weather station.[9]

Wegener grew increasingly anxious as delays began to multiply. Summer was slipping away and Eismitte was not yet established. The late spring melting of sea ice and the possibility of winter storms beginning as early as September added more urgency to his efforts. Georgi's sled trip to Eismitte started on July 15, later than expected, and he took a load of only 680 kilograms (1,500 pounds). If three men were to spend the winter at Eismitte about 3,200 kilograms (7,000 pounds) of provisions and equipment must be transported there—five more trips of that size. The two propeller-driven motor sleds had not yet been assembled and made operational. The mechanics assembling the motor sleds reported they would probably not be ready before the tenth of August. They would need shakedown runs to discover their capabilities in different snow conditions, on uphill slopes, and under various loads. The sleds were crucial to the plan as dogsleds alone would not be able to carry the necessary provisions for Eismitte in the short time available.

The trip to Eismitte by dogsled took fourteen days in good weather and snow surface conditions and about half that for the less strenuous return trip. Each round-trip would require at least three weeks—often more—of the limited time available. Only by having many more dogs and dogsleds could the transfer of goods to Eismitte be done without the motor sleds.

Several unmet expectations increased Wegener's disappointment in the expedition thus far. They had been almost six weeks late in reaching Kamarujuk Fjord because of sea ice not breaking up as soon as expected. The transport of all the provisions and equipment up to West Station took longer than planned. The assembly and testing of the motor sleds was moving more slowly than anticipated. And it was still uncertain if the motor sleds would be able to carry as much or move as fast as expected. These worries began to appear in his logbook almost daily. The thought of failure began to plague him. The possibility of getting winter meteorological data from the center of Greenland had been a major selling point for the project, and Wegener's sense of responsibility for the project's success weighed heavily on him.

On August 5 Wegener found a louse in his shirt. This unpleasant discovery made him shudder in disgust. He attributed the presence of the louse to living in such close contact with the Greenlanders who were infested with them. But through all the troubles, Wegener never became visibly angry or lost his self-control. He used his diary as a buffer against his frustrations. On this occasion he wrote:

> Today I'm feeling depressed and pessimistic. Success seems to be slipping from our grasp because our means of transportation are not effective enough. Or is it merely the louse this morning? I was busy the whole day washing, boiling clothes and bedding, cleaning things with gasoline and spraying with insect spray, and have done nothing to help on the good cause.
>
> At midnight now the sun is already so far below the horizon that it becomes noticeably darker. The winter night is casting its shadow, and we shall be ill equipped to meet it. If winter sets in early we shall fall right on our faces.[10]

A few days later they discovered that there was only enough hay to feed the ponies for twenty more days. Something had to be done quickly as there was still much work for the ponies. Almost no grass grew in the area, so they could not make hay.

To add to the gloom, some of the men developed ailments. One had a case of "nerves"; Lissey had attacks of sciatica resulting from a bruise; Wölken was exhausted and had to be released from work; Vigfus had rheumatism. Jon, an Icelander, may have had an ulcer. He got sick when he drank too much coffee but he would not give up coffee. Unfortunately, no one on the expedition was a physician. Wegener wrote of his growing concern: "So the crisis is still on and getting more intense. The expedition's prospects are clouded over. We are not going to deceive ourselves and go on working as if all were going well. My companions have been grossly overworked the whole time, and their energies are beginning to fail. This may be our greatest difficulty. How will it all end? This is now a burning question. Since I found the louse the difficulties have grown in a very worrying way. Since I found the louse! Before anything else I must get some proper sleep. We will not give up lightly."[11]

The next day Wegener took the *Krabbe* to Umanak and met with the local town leader. They had hay available and would bag about four hundred pounds for the expedition. This was not a significant amount of hay, but it would help relieve their desperate situation. Wegener also arranged for additional sled dogs and some men to manage them. The next day the Greenlanders delivered and loaded the hay onto the boat. As the *Krabbe* began its return to Kamarujuk Fjord, it carried four hundred pounds of sacked hay, seven Greenlanders, and 40 dogs. Wegener complained that the dogs were filthy and stank like the plague. All together the expedition now had 143 dogs, mostly borrowed, and Wegener was thinking of getting 20 more.

Back at Kamarujuk Wegener found that the rate of supplies transport had increased since they began using the moraine path regularly. Also he found another source for seven tons of grass that still needed to be dried. Things were taking a turn for the better. Now they had fodder to spare, and the mood of the camp improved. As of August 20 only a few remnants of goods remained at the storage depot in Kamarujuk Fjord at the foot of the glacier. All the rest was in the new storage place atop the glacier at the West Station. The winter hut was ready for assembly. But their worries were not over. Jon's condition worsened and he now vomited blood. The *Krabbe*'s engine began misfiring and losing power. However, that proved to be only a loose bolt in the governor. Although someone was able to fix it, a few days were lost searching for the problem.

The additional hay arrived and had to be removed from the bags immediately as it was not properly dried. The grass had begun to rot in the bags and had become matted into solid masses. Salvaging it required pulling the clumps completely apart and spreading the grass on the ground to dry. Fortunately, the ponies thought all the dried hay was edible, so no serious loss from rot had occurred.

The optimism of the camp began to wane as the last week of August arrived. Greenland summer was nearly ended. The second sled party, just returning from taking a load of provisions to Eismitte, announced that the temperature there was already down to -35° Celsius (-31° Fahrenheit). Their greatest source of optimism came with the announcement that the motor sleds were finally assembled and the engines were running for the first time. The sleds still required some testing, but the expedition members felt that the rate of transport to Eismitte would at last begin to move more rapidly.

Almost everything necessary for the West Station and for the Eismitte was now stockpiled at West Station and ready for assembly or for transport to Eismitte. The arduous task of moving tons of goods up the glacier was finally over. The winter hut was ready for assembly at West Station. Even the gasoline storage site had been moved to the top of the glacier. Had the West Station been the only site to be established, they would have been in fine shape.

Establishing bases on the west and east coasts required weeks of hard labor, but proved to be manageable given sufficient manpower. However, moving four tons of equipment and provisions 400 kilometers (250 miles) from West Station to a mid-ice station depended on much more than hard work. The unpredictable weather determined whether future dogsled trips were possible and whether they would take more than the ideal fourteen days to reach Eismitte.

# 6 ❖

## Establishing Eismitte

*As courage and intelligence are the two qualifications best worth*

*a good man's cultivation, so it is the first part of intelligence to*

*recognize our precarious estate in life, and the first part of courage*

*to be not at all abashed before the fact.*

—ROBERT LOUIS STEVENSON

Creating a meteorological station in the middle of the Greenland ice cap changed a fairly simple expedition into an extremely difficult one. Distance, weather, and altitude created one hurdle after another for the Central Station team. Transporting equipment and provisions 400 kilometers (250 miles) by dogsled usually took two weeks even in the best weather conditions. Averaging 29 kilometers (18 miles) per day demanded great energy from men and dogs. And the loads for the weather station were limited because of food, tents, and bedding needed for the trip out and back. Storage depots were established along the way, but these were held in reserve for emergency use in case a sled party was stranded by a storm. Another hurdle was the expedition's dependence on motor sleds that had not been tested in the Arctic. As the team realized the limits on loads carried by dogsleds, they realized that motor sleds were more than a supplement to the dogsleds—they were essential.

The elevation of the ice cap increased from 915 meters (3,000 feet) at West Station to 3,003 meters (9,850 feet) at Central Station. The gradual rise was the most subtle hurdle of all. Though it was hardly perceptible, as days passed the men's energy was sapped by the thinning atmosphere as well as ordinary fatigue from a difficult journey.

During the Greenland reconnaissance trip in 1929, Wegener, Loewe, and Georgi made two trips by dogsled half the distance toward the site that would become Central Station (Eismitte). From this experience they estimated that three dogsled trips with twenty sleds each could transport 3,180 kilograms (7,000 pounds), and any additional supplies needed could be brought by the motor sleds. Soon the estimate of provisions needed for Eismitte increased to 3,640 kilograms (8,000 pounds) including a prefabricated winter hut to be assembled on site and additional kerosene for cooking and heating. The need for extra supplies increased their dependence on the motor sleds as the onset of winter would bring weather too severe for any travel.

The first sled trip from West Station to establish Eismitte began in mid-July 1930. The eleven-sled party included Johannes Georgi, Fritz Loewe, Karl Weiken, and eight Greenlanders. Weiken was being trained in the art of dogsled driving by one of the Greenlanders. Georgi and Loewe were already able to drive their own sleds.

The Greenlanders thought the planned load for the sleds was too heavy for such a long trip and insisted on cutting the equipment weight by 230 kilograms (500 pounds). This made a total of 955 kilograms (2,100) pounds, and the weight was apportioned according to the number of dogs on each sled. If they made three trips like this there would be only a little over 2,730 kilograms (6,000 pounds) of provisions and equipment at Eismitte by winter. The Greenlanders still thought the weight was too great, but they agreed to go anyway.

The trip began at 5 P.M. on July 14 in a misty rain mixed with snow. By 11:30 P.M. they had gone 20 kilometers (12.5 miles) in six and a half hours—only 3 kilometers per hour or just under 2 miles per hour. They made camp in a heavy rain that lasted into the middle of the next day. Now everything was wet, and the contents of the uncovered boxes were soaked. That evening they started off again, knowing that daylight would last all night. The visibility was poor, the snow was slushy, and progress was slow. In another 20 kilometers the dogs were exhausted. By now they knew that they needed more rest and more food to make such a trip. The Greenlanders began to complain that they were not getting enough to eat. More pemmican was added to the soup to strengthen it, and the dogs' rations of pemmican were increased too. As they progressed the weather became colder and the snow firmer. This made travel easier, and they began to make 25 kilometers (15.5 miles) per day. On the fifth day of travel they passed the 100 kilometer (62 miles) marker—fair progress, but not as good as they thought necessary.

The trackless and featureless surface of the ice cap required some sort of guidepost markers to keep future travelers on the path. They marked the route with black flags on poles every 500 meters (0.3 miles), and with a cairn built of snow every 5 kilometers (3 miles). They measured distance by attaching an odometer the size of a bicycle wheel to one of the sleds (figure 6.1). This sled party, guided by compass, installed the route markers as they went, and at each cairn they attached a large piece of black cloth held out in a rectangle by poles stuck in the top of the cairn. This method of trail marking generally worked well, but even a 500-meter interval between flags seems long when visibility is low from fog or blowing snow.

The Greenland ice cap was a fearsome and forbidding place to the local Inuits who traveled only in the coastal areas and on sea ice. The main problem for them was the lack of landmarks to guide them back home if they ventured too far onto the ice cap. Normally they moved overland only when crossing a peninsula from one fjord to another. This reluctance to travel on the ice cap

FIGURE 6.1
On trips to Eismitte, an odometer wheel (*left side of sled*) was attached to one sled to keep an accurate measurement of distance traveled. (Used with permission of the Alfred Wegener Institute for Polar and Marine Research, Bremerhaven.)

created problems for Wegener's expedition, and the Germans frequently had to persuade the Greenlanders to continue, especially when they encountered fog and blowing snow.

Finally the Greenlanders conferred among themselves and announced it was time to turn back to the coast. Georgi and Loewe determined which Greenlanders were willing to go all the way to Eismitte and which would agree to go only as far as the halfway marker. This was a serious dilemma. They had started with a lighter cargo than planned; if they now had to cache much of it at the halfway marker, provisioning Eismitte would take even longer than expected. The Germans separated the two groups of Greenlanders in hopes of keeping the reluctant ones from influencing the willing ones. The Germans stayed with the willing group day and night to keep them from talking to the others, and they traveled in separate parties.

Despite all these precautions, when the sled party reached the 200-kilometer (125-mile) marker on July 22, all the Greenlanders announced that they wanted to return home immediately. The Greenlanders were afraid they might not be able to breathe if they went farther, or the dogs might collapse, forcing the men to walk, which meant they might die of hunger. The three Germans could not manage all the sleds themselves, so a long session of persuasion and negotiation took place. Finally the group that earlier had been willing to go the full distance, reconsidered and agreed to continue. Loewe agreed to return to West Station with the other group as they had no experience at navigation on the trackless ice cap and wanted guidance.

Three Greenlanders agreed to continue with Georgi and Weiken toward Eismitte provided the sled loads were reduced. This was agreed to, and provisions were kept at the expense of equipment, which was stowed in the open at the halfway point. This arrangement was a blow to the expedition's plan but was still better than having to turn back and give up establishing the Central Station. On July 30 they reached the site for Eismitte, ending the first dogsled trip with a sense of accomplishment tempered by disappointment at its diminished scale. They had reached the chosen point for Eismitte, had marked the entire route along the way, and had gained the trust of Greenlanders for future trips by showing that the ice cap could be traveled without getting lost.

On July 31 they began setting up a weather station at Eismitte with thermometers and a barometer in the standard ventilated box mounted on an ice pedestal. The next day Weiken and the three Greenlanders began the return trip to West Station. Georgi stayed alone at Eismitte with a tent for shelter and began the routine of recording meteorological data. The weather was bright and calm, and the tent would provide a warm shelter as long as the sun continued its twenty-four-hour radiation. Loewe was expected in two weeks with more supplies, including some items from the cache at the 200-kilometer depot.

Georgi's first task was to dig a subsurface room. The barometer would be put there for better protection, and Georgi could begin measuring ice temperatures below the surface. Georgi quickly discovered that the nighttime temperatures caused the clock-driven recording instruments to stop working at -23° Celsius (-9° Fahrenheit). He was able to make modifications and get them working again, but this problem of the contrary clocks reappeared several times through the winter.

Each time Georgi went outside he looked expectantly to the west hoping to see the second sled party on the horizon. On August 18 he finally saw a dot on the horizon. An hour and a half later Loewe and five Greenlanders arrived with 910 kilograms (2,000 pounds) of supplies, equipment, and mail for Georgi. By now Georgi had been alone at Eismitte for eighteen days, and the thought of some friendly company lifted his spirits. Eismitte was progressing, but the accumulation of supplies was far below the original expectations.

On August 19 Loewe and his party departed for West Station, and Georgi was again alone at Eismitte. After a few hours Rasmus Villumsen, one of the Inuits, returned to Eismitte to pick up Loewe's sleeping bag, which had been left behind. If that loss had not been discovered until night fall, it would have been a disaster for Loewe.

During his time alone, Georgi was fully occupied with improving living conditions at Eismitte and recording data. He began the first observations of the day by measuring temperature, atmospheric pressure, wind, and humidity at 7:40 A.M. Then he returned to the tent to cook breakfast and spent the morning working on clocks, camera shutters, and wind recorders. Almost every instrument needed some modification before it would work at ice cap temperatures. The rest of the time he spent digging the ice cave. By 9 P.M. the temperature dropped to -29° Celsius (-20° Fahrenheit) and Georgi quickly ate a liter of pemmican soup and crawled into his reindeer sleeping bag.

On the thirteenth of September the third sled party arrived at Eismitte with three Germans (Sorge, Wölken, and Jülg), seven Greenlanders, and 1,360 kilograms (3,000 pounds) of supplies. This load was big enough to bring the supply of provisions, fuel, and equipment to a more promising level, but the original plan did not include more dogsled trips this late in the season. Winter storms could start at any time. The motor sleds were expected to finish the job of transporting four tons of supplies. Still missing from the stockpile at Eismitte was a sufficient quantity of kerosene, explosives for seismic sounding of ice thickness, the winter hut, a radio, and certain instruments. The motor sleds remained at West Station being assembled and tested for travel. With luck there might still be time for another dogsled trip in October if winter did not start early.

Georgi and Loewe made a list to send to Wegener back at the West Station naming priority items needed in case only one more sled trip could be made. They decided to settle for twenty-seven canisters (40 liters [10.5 gallons] each) of kerosene rather than forty-six as had been planned. This much kerosene plus the winter hut would make a full load for ten dogsleds even omitting the radio and more scientific equipment.

Having a radio held a surprisingly low priority in their minds, though they held frequent discussions about it. No one on the expedition had experience with radios in Arctic winters, but they knew that instruments in general behaved erratically at low temperatures. The conclusion was that it would be better to have no radio than to have one that suddenly quit working. Loss of communication might lead to an assumption by the people at West Station that something serious had happened and cause them to launch a rescue expedition in the middle of winter. This could endanger the lives of the entire rescue party. However, it would seem that with some advance understanding on procedures to follow in case of communications failure, they might have moved the radio to higher priority. Radios in 1930 were quite heavy, and weight might have been a factor in their consideration of a radio for Eismitte, particularly when choosing items to reduce the load on the sleds. They had a radio on the last sled trip to Eismitte, but it was cached along the way. Top priorities on the list of items for Eismitte were fuel, food, explosives for seismic depth soundings, and the most basic meteorological instruments.[1]

The message Georgi sent to Wegener at West Station arrived in time for another sled party to reach Eismitte by the middle of October. Georgi also said that if the fourth sled party did not arrive by October 20 that he and Sorge would leave Eismitte and start back for West Station pulling hand sleds carrying food, tents, and sleeping bags. They felt that if the additional provisions did not arrive by the twentieth of October, they should not risk staying at Eismitte for the winter with low fuel and food supplies. If the additional sled party did not come they would have only one-third of the absolute minimum of kerosene needed. The deadline they selected was somewhat arbitrary, but they needed to allow enough time for their safe return to West Station. Pulling hand sleds was slower and more tiring than traveling by dogsleds, but they thought the risk of foot travel was less than the risk of wintering at Eismitte with insufficient fuel.

From Central Ice Station
September 14, 1930

Dear Wegener:

We reached the ice station yesterday and are all as fit and cheery as possible. We unloaded here in all 1,530 kilograms [3,370 pounds]. The food supplies here would last two men for 10–11 months, three men for seven months. If three men are to winter here, please send us three red boxes and the two red extra boxes at a minimum. The fuel supply is not nearly sufficient. If we use 3.3 liters [7 pints] a day, one can will last ten days, three cans a month, 27 cans in nine months. So we need 17 cans of fuel (weight 682 kg) [1,500 lb].

The fact that the motor sleds have not arrived creates a new and dangerous situation for us. We two agree to winter here together, even if the hut cannot be brought here as a consequence of the failure of the motor sleds, but only on the following assumption: by October 20 (the latest date for returning) we must have 17 large cans of fuel, drilling tools, snow bucket and rucksack with contents.

The letter continues with a list of other items needed including explosives and equipment for seismic work. Then it concludes:

If these necessities are not here, or their arrival definitely announced, by October 20, we shall start on that day with hand sleds. But we hope that you will arrive any day with motor sleds, and that all difficulties will be thus removed.

With best regards to all,
Georgi, Sorge[2]

On September 13 Jülg, Wölken, and the Greenlanders left for West Station. Georgi and Sorge, still at Eismitte, continued digging out the subsurface rooms while they waited for the fourth sled party to arrive. In the first week of October the weather began to change, and Georgi was concerned that the third sled party might still be en route back to the West Station. On the fifth of October the temperature dropped to -40° Celsius (-40° Fahrenheit) and by the tenth of October it was down to -46° Celsius (-50° Fahrenheit). They now realized that winter was starting early rather than late as they had hoped. They also knew that no sled party should try to reach them in weather like this, and they began making trial runs with the hand sled in preparation for a trip to West Station on foot. They found the hand sled runners needed to be longer for the load they planned to carry so they attached their only pair of skis for runners on the sled. This meant they would both have to snowshoe to the West Station pulling a sled. Georgi had traveled in this manner during the 1929 reconnaissance expedition and knew it was possible, though more tiring than with dogsleds.

As Georgi and Sorge settled into the ice cave, they were aware that it was fairly comfortable and easy to heat if ventilation were kept to a minimum. Through trial and error with the ventilation cover they learned to achieve the optimum temperature that would not melt the roof, yet still provide enough air flow to remove fumes through the vent. The winter hut would have been warmer, but would have used much more fuel. They made some new calculations on fuel needs and concluded they could get by on much less kerosene than they previously thought. When the third sled party left for the West Station, Georgi thought staying on at Eismitte without more supplies was out of the question. Now they felt there was a reasonable possibility they could get through the winter with careful conservation of food and fuel. When the October 20 deadline arrived they decided not to abandon Eismitte even if no additional supplies came. By now the risk of traveling to West Station on foot in winter blizzards appeared greater than the risk of staying at Eismitte. Wegener certainly would have agreed with their logic and would have approved of this plan. Unfortunately, they had no way to tell him of their decision.

Their change in plans was motivated by their commitment to the expedition's goals and strong loyalty to the leader, Alfred Wegener. Abandoning Eismitte would be a crushing blow to the central goal of the project—creating a yearlong analysis of the atmosphere across Greenland. Although they knew Wegener would accept their decision to leave if they felt it necessary, they did not want to let him down. Everyone on the expedition was committed to the project and held Wegener in high regard. He was such a congenial personality and devoted scientist that each man wanted to give his utmost to the cause. Georgi and Sorge were also committed scientists and had a personal stake in the success of the expedition. Furthermore, they knew that Wegener placed a high value on independent decisions made by senior members of the expedition and would honor whatever they decided to do. Wegener had often expressed the view that men on the scene should be able to alter plans when conditions were different from those anticipated.

## Testing the Motor Sleds

On August 29 the two motor sleds were ready for testing at West Station. With a barrage of snowballs, the group christened the sleds Schneespatz (Snow Sparrow) and Eisbär (Polar Bear). They had high expectations for using the sleds as transport. Riders would be sheltered from the wind, and the expected speed was much greater than dogsleds. The fuselage of the sleds was designed to hold two men along with about 400 kilograms (880 pounds) of cargo, which included tents and provisions for the crew. A steering wheel turned the front runners.

That same day they set off on the first trip with full loads. Wegener and Schif were in the Schneespatz, Kelbl and Kraus took the Eisbär. Soon they began to have problems. Every time they stopped, the runners froze in place and had to be chipped loose. This was solved by putting boards under the runners while stopped. The route marker flags were missing for the first 24 kilometers (15 miles), and the party was afraid of getting lost. They tried guiding by compass and estimating distance to find the first fuel depot. This type of navigation was unreliable because the slopes had to be traversed by a zigzag movement. The sleds had insufficient power to push directly through drifted snow in an upslope direction. Also, strong headwinds and wet snow further hindered their efforts. As daylight faded they decided to give up and turn back for West Station.

On the second try, September 2, the weather was fair and calm. They went only as far as the 24-kilometer (15-mile) depot and returned, replacing the missing markers on the return trip. They now could see that progress in the motor sleds depended on the wind, especially when traveling east, which was upslope. A third attempt failed because of bad weather.

The sleds were tested a fourth time on September 5. Kraus, Kelbl, Schif, and one Greenlander made good progress after 48 kilometers (30 miles). Before that point their progress was little more than a fast walk. Beyond that the eastward rise of the glacier surface flattened a bit, the snow became drier, and they were able to make about 24 kilometers per hour (15 mph). They found that the sled engines were not strong enough to operate except in the best conditions—calm winds, dry snow, and gentle slopes. This made travel with sleds in foul weather unfeasible. The sled drivers concluded that they should travel only in good weather and at a reduced speed.

They devoted this fourth motor sled trip to establishing fuel and provision depots along the route to Eismitte. Each sled made trips in stages to the halfway station over a period of a week. The goods accumulated there included kerosene for stoves, gasoline for motor sleds, the winter hut for Eismitte, instruments, and household utensils. Altogether they cached about 1,230 kilograms (2,700 pounds) at the halfway marker.

During this work, they discovered several problems with sled operations. In the low temperatures, fuel lines and carburetor jets froze and engines were hard to start. A fuel leak occurred and lines had to be repaired with materials like string, tape, and wire. The thought of a fuel leak far from a fuel depot was disturbing, conjuring up the possibility of being stranded with a minimum of provisions.

On September 17 they again left West Station for Eismitte and reached the midpoint depot just as darkness began. At this point they met Jülg and Wölken returning from the third trip to Eismitte by dogsled. They all camped together at the 200-kilometer (125-mile) station, and felt optimistic about reaching Eismitte the next day with another supply of provisions and fuel. They planned to shuttle the motor sleds back and forth from the midpoint depot to Eismitte until all the fuel and provisions had been delivered.

The next morning brought a driving snow with low visibility—the most dangerous condition for travel. Without being able to see the route markers the sled parties would have quickly become disoriented. The group with the motor sleds decided to wait for a break in the weather. The returning dogsleds continued westward in the bad weather and arrived at the West Station without trouble.

As the day progressed the weather worsened. In two hours the tents were drifted over, and the motor sled crews had to dig themselves out. After three days, their tents, clothes, and sleeping bags were damp from the lack of ventilation in the tents and the accumulation of condensation. Late in the third day the storm began to subside and the sky showed signs of clearing. They hoped to resume their trip the next morning.

They spent the entire next day heating the engines trying to get them started. Then another day was spent digging out the sleds' frozen runners and trying to make them move. By the end of the day they were all exhausted from the heavy exertion at 2,500 meters (8,200 feet). On the sixth day snow and fog again reduced visibility too much for safe travel. It appeared that winter storms had come to stay. Provisions for the sled party were getting dangerously low, and they had already used some of the emergency rations that had been cached at the halfway marker.

On the seventh day the storm diminished, and they decided to try traveling again. They managed to start the engines and get the sleds to move. Unfortunately, deep fresh snow and strong winds immobilized the heavily loaded motor sleds. Even under full power with a person outside pushing, the sleds would not move. They felt they had no choice at this point but to give up their attempt to get the motor sleds to Eismitte. This decision was one of the severest blows to the expedition. They were only a few hours from Eismitte. They unloaded and cached all the provisions, equipment, and instruments intended for Eismitte. Perhaps they should have attempted to carry a lightened load to Eismitte.

Now the motor sleds began the return to West Station with only the crew and meager provisions for the trip. Just 39 kilometers (24 miles) from their destination the engine of the Schneespatz overheated and stopped. They stowed the sled, and all four men loaded into the Eisbär—including a disabled dog left behind by the westbound third dogsled party. Even though it was getting dark they could expect to reach West Station in less than two hours. Their plan was to return the next day with tools to repair Schneespatz. After only 8 kilometers (5 miles), Eisbär also overheated. They pitched tents and waited for morning. They had no food but pemmican, and their tobacco was gone. They faced a miserable night.

The next morning Wegener and Loewe appeared with fifteen sleds and thirteen Greenlanders headed for Eismitte on the fourth dogsled trip. They had camped for the night only 3 kilometers (2 miles) from the motor sled crew and had seen lights from their camp in the darkness. Wegener was a welcome sight as his party had sufficient food to supply the motor sled group with a decent breakfast and some tobacco.

They told Wegener about the failure of the motor sleds. Realizing that Eismitte would not have adequate supplies to support two men for the winter, Wegener decided his best option was to reach the halfway point and reload his sleds with cached items of the highest priority. Kurt Schif wrote that Wegener seemed unsurprised that the motor sleds had not performed as expected. He seemed almost to have expected them to fail. Wegener may have concealed his disappointment from the group, but perhaps he had become disillusioned by the repeated delays to that point and had lowered all his expectations. This would be an alarming change of heart for a man with such a positive outlook as Wegener.

The motor sled crew managed to start Eisbär, but in a few minutes the sled had broken down again. They had to give up all hope of getting the sled back to West Station this winter. They abandoned the sled and continued on foot. They got out an emergency hand-pulled sled, their Arctic lifeboat, which was carried on the motor sleds. They loaded a tent, sleeping bags, and a Primus stove into the wicker sled and headed for West Station 31 kilometers (19 miles) away. As night fell they neared the crevasse zone and stopped to avoid the dangerous crossing in the dark. By now their food supply consisted of two pieces of hard bread, a few lumps of sugar, and oddly, a piece of chocolate cake given to them by Wegener.

The next morning another heavy snowstorm was blowing, and all route markers were invisible. They moved ahead anyway using a compass. Suddenly Kurt Schif fell into a crevasse that had been covered over with crusted snow. It was not deep, and they got him out quickly, but it made them realize their desperate situation. They pitched a tent to wait out the weather. Four very depressed and hungry men and one dog huddled in a single emergency tent only 30 kilometers (18 miles) from their destination with no food or fuel and a blizzard blowing outside.

On the third day the blizzard stopped, and they could get their bearings by the sight of mountains at the edge of the ice cap. They roped themselves together for protection from the crevasses and resumed the trip westward. They were hungry, their energy was low, and progress was slow, but they finally made it to West Station on September 27. The two motor sleds were abandoned for the rest of the winter. The sleds had played a major role in the planning for transport of goods to Eismitte, but they were not suited for the winter storms that confronted them. What had begun with high hopes ended in failure.

# 7 ❖

## The Fourth Trip to Eismitte, September 1930

*The brave man is not he who feels no fear, for that were stupid and*

*irrational; but he whose noble soul subdues its fear and bravely*

*dares the danger nature shrinks from.*

—JACQUES MARITAIN

By the middle of September, the season favorable for provisioning Eismitte was nearly over. Three trips by dogsled had delivered only part of the support needed for two men to spend the winter in the middle of Greenland. The third sled party had just returned. Wegener had not yet seen the note from Georgi saying they would abandon Eismitte and return by hand-pulled sleds if the fourth party had not arrived with more supplies by October 20. Nor did he know that the motor sleds had failed to reach Eismitte and lay abandoned a few miles from West Station. (Table 7.1 shows the dates of the trips made to and from Eismitte.)

With these unknowns facing him, Wegener decided a fourth dogsled trip to the mid-ice station with at least fifteen sleds must begin at once. This trip would ensure enough winter supplies for Georgi and Sorge. They could not leave immediately because the third sled party was not expected before September 20. Until they returned there were not enough sleds or dogs for a fourth trip. Moreover, the sleds still out on the third expedition were of a broad runner construction needed on the softer snow of the ice cap surface. The narrower runners of most Greenland sleds were fine for icy and stony surfaces but cut too deeply into the snow on the ice cap.

TABLE 7.1 *Chronology of Wegener's fourth expedition*

| Date | Activity |
|------|----------|
| 1930 | |
| April 1 | Depart from Copenhagen |
| May 6 | Unload ship on pack ice and transport gear to Uvkusigsat |
| June 18 | Transport gear from Uvkusigsat to Kamarujuk glacier |
| June 18 | Begin transport up the Kamarujuk glacier |
| July 14 | Begin first trip to Eismitte (Georgi, Loewe, Weiken, and 8 Greenlanders) |
| July 30 | First party reaches site for Eismitte |
| August 6 | First party returns from Eismitte (Georgi remains behind) |
| August 6 | Begin second trip to Eismitte (Loewe and 5 Greenlanders) |
| August 18 | Second party arrives at Eismitte |
| August 25 | Second party returns from Eismitte |
| August 29 | First test of motor sleds |
| August 30 | Begin third trip to Eismitte (Sorge, Jülg, Lölken, and 7 Greenlanders) |
| September 13 | Third party arrives at Eismitte |
| September 21 | Third party returns from Eismitte (Jülg, Wölken, and Greenlanders; Sorge remains with Georgi) |
| September 22 | Begin fouth trip to Eismitte (Wegener, Loewe, and 13 Greenlanders) |
| September 27 | Motor sleds abandoned (Kraus, Kelbl, Schif, and 1 Greenlander) |
| October 29 | Fourth party arrives at Eismitte (Wegener, Loewe, and Villumsen) |
| November 1 | Wegener and Villumsen leave Eismitte (Loewe remains behind) |
| 1931 | |
| May 7 | Spring relief party arrives at Eismitte |
| August 7 | Eismitte abandoned |

Wegener also needed to find more dogs. For a large party with fifteen sleds he would need about 130 dogs with 8 or 9 dogs per sled. Greenlanders in the nearby villages agreed to rent their dogs for five kroner per dog, with a payment of fifteen kroner for any dog that died on the trip. The Greenlanders who came along to drive the dogsleds were paid three kroner per day while in camp and four kroner per day while traveling—more if conditions were particularly hard. The expedition would furnish food for men and dogs.

Wegener made the decision to lead the fourth expedition himself in place of Karl Weiken. He could see that some important decisions would have to be made regarding supplies, and Wegener wanted to make those decisions himself. Weiken and other members of the expedition proposed that they begin without waiting for the return of all the sleds. In hindsight, it might have been a good decision, a way to beat the storms and reach Eismitte with some supplies.

On September 19 and 20 members of the expedition assembled all supplies, sleds, and dogs at West Station in preparation for departure as soon as the third sled party arrived from Eismitte. In the early morning hours of the twenty-first, all one hundred dogs at West Station began to howl for no apparent reason. The howling not only interrupted sleep for everyone in the camp but also created an eerie sense of foreboding. Each man must have felt the hairs on his neck stiffen as the dogs wailed.

Preparations for the trip included covering the dogs feet with *kamiks*, sacks of sailcloth or sealskin tied to their legs with holes cut for their claws. Sharp ice crystals about 10 to 15 millimeters (about 0.5 inches) long on the surface of the glacier cut the dog's paws, especially when pulling heavy loads up a slope where the paws bore down harder. The kamiks had to be removed during any prolonged stops so the dogs could lick their paws and because the dogs would eat the kamiks if given the chance.

On the morning of September 21 the third sled party with Jülg, Wölken, and seven Greenlanders arrived at West Station, having departed Eismitte on August 30. They gave Wegener the letter from Georgi and Sorge telling what supplies were of highest priority for the fourth party to bring to Eismitte. It was fortunate that they arrived before Wegener and the fourth sled party departed, as there were a number of items requested in the letter that had not been included in the load for Eismitte. Some items in reserve had to be brought up to West Station from Kamarujuk down on the coast, so it took an extra day before the fourth sled party was ready to go. On September 22 they started for Eismitte and covered 17 kilometers (10.5 miles) before stopping for the night.

They camped the first night near the stalled motor sleds. While there they decided to lighten the dogsled loads by a total of 730 kilograms (1,600 pounds) which they stored at the site with the motor sleds. They hoped they could make faster progress with the lighter loads. Because they had cached the equivalent of one sled load, they were able to send one Greenlander and sled back to West Station. As they continued the next morning they could see low stratus clouds signaling the approach of a storm.

Progress was painfully slow for the first few days because of the fresh snow and heavy loads. On dogsled trips the load on each sled will lighten as much as 5.5 kilograms (12 pounds) per day due to consumption of food by the dogs and the driver. As the loads decreased, progress became easier. Unlike some polar expeditions, Wegener's group did not plan on killing unneeded dogs as the load lightened and using dog meat to feed the remaining dogs. This practice would have allowed the sleds to carry a smaller amount of dog food. By 50 kilometers (31 miles) out from West Station they were traveling in a thick fog and heavy snow which continued for two days, covering sleds, tents, and dogs during the nights. By the fifth day of travel the sky was wonderfully clear with a strong cold wind and drifting snow. At night the temperature dropped to -27° Celsius (-17° Fahrenheit).

That night all the Greenlanders gathered in one tent, a sign that they were having a decision-making conference. In the morning the Greenlanders came to the Germans' tent and sat down silently smoking their pipes. After a period of silence the spokesman announced that they all wanted to go home. They did not have enough clothing to keep out the cold, and they had already begun to suffer. Also, they felt their sleds were still too heavy for the soft condition of the snow. The German team wore the same native clothing and may have mistrusted the Greenlanders' reasons, nevertheless, this was a major crisis for the expedition. Though hampered by language difficulties, the Germans tried to change the Greenlanders' minds. Even Wegener, whom the Greenlanders admired, could not dissuade them.

The Greenlanders' typical winter clothing comes primarily from reindeer and dog skins, and certain skins are considered more suitable for winter weather. Clothing and sleeping bags made from reindeer skins are valued because the multichambered hair cells provide excellent lightweight insulation.[1] Inuit clothing makers prefer skins of adult cows killed early in the autumn, before the thickening of the winter coats, to provide the maximum warmth and the lightest weight. Late fall skins from either cows or bulls are too heavy for comfortable clothing but make good bedding. Calfskins may be used for boot linings. The skins of reindeer forelegs make boot uppers and mitten palms because they are resistant to abrasion. Trousers are often made of dog skins. Wolverine or wolf fur is used for the ruff around the edge of the parka hood because those furs shed the ice crystals that form from breathing. Native clothing was somewhat fragile and required frequent repairs, drying, and softening, but was warmer and lighter than any Western clothing the expedition party brought.

Loewe reported in his notes that he couldn't help but agree with the plight of the Greenlanders. Even though the expedition had provided some supplies to the Greenlanders, most of them had no reindeer-skin sleeping bags. The party needed more equipment for such cold weather. They were about to be beaten by the weather. After some further discussion with the Germans, four Greenlanders agreed to continue toward Eismitte for an increase in pay.

After starting with fifteen sleds, they were now reduced to six sleds, two Germans, and four Greenlanders. This meant losing a day to remove and repack everything and reduce the total load by much more than they had before. The excess load was stored along the trail near one of the snow cairns, and its location noted in the logbook. On September 29 they resumed the trip to Eismitte with sixty-nine dogs and barely 1,800 kilograms (4,000 pounds) of supplies for Eismitte.

As Wegener's group continued toward Eismitte, the weather was ideal for travel, with daytime temperatures of -10 degrees Celsius (14 degrees Fahrenheit) and light winds. In spite of some light fog they were able to find all the route markers even though many of them were gradually becoming obscured by snow drifts. They rebuilt the snow cairns as they went and raised the flags higher to make their return trip safer. On October 1 they reached the 120-kilometer (75-mile) cairn. Again, the four Greenlanders wanted to turn back and head for home, but they were persuaded to continue. The next morning the group could not travel because of heavy fog and snow, so they had to lay over for the day. During this break they decided to lighten the load again and leave some items at the 120-kilometer marker.

The most pressing item to take to Eismitte was kerosene for heating and cooking, and it had been stored earlier by the motor sleds at their greatest distance of travel, the 200-kilometer (125-mile) storage depot. So on the third of October they started again with almost no load except for the supplies the sled party itself needed for travel. They planned to retrieve the kerosene cached by the motor sleds and take it on to Eismitte.

For several days the weather was relatively mild, but travel was slow due to the deep, fresh snow. They slowed to only 15 kilometers (9 miles) per day compared with summer travel of up to 35 kilometers (22 miles) per day even with much heavier loads. With such slow progress they realized they did not have enough food for the party of six to reach Eismitte.

After lengthy discussions, Wegener and Loewe decided they should give up the trip if conditions were not improved by October 6. On the evening of October 5, the exhausted Greenlander who had been breaking trail through the deep snow announced that the Greenlanders were definitely leaving. This caused another lost day of good travel weather, but negotiations resulted in the Greenlanders agreeing to continue to the 200-kilometer (125-mile) site. However, they soon changed their minds and said they would go only to the 150-kilometer (93-mile) point, and then they were going home.

None of the Greenlanders knew any European language, and Wegener and Loewe spoke only a smattering of Inuit. At one point they unpacked and reloaded, thinking they had a final agreement, only to have one of the Green-landers again change his mind. After several arduous, time-consuming, and tiresome attempts at negotiation, one Greenlander, Rasmus Villumsen, agreed to stay with the party all the way to Eismitte. Villumsen, having traveled with Loewe to the interior, may have had more confidence about traveling in the unknown territory than the other Greenlanders. Before proceeding they had to open boxes, take out the necessary items for the rest of the trip, and weigh and repack them.

On October 7 Wegener, Loewe, and Rasmus Villumsen parted company with the three Greenlanders who were returning to West Station and began moving eastward again. They felt a great sense of urgency, believing they needed to reach Eismitte before October 20 when Georgi and Sorge had said they would leave the Central Station. They did not know that Georgi and Sorge had changed their minds and would be wintering at Eismitte in any case. The returning group of Greenlanders reached West Station without inci-dent on October 15.

Eastward progress continued slowly in soft, belly-deep powder through which the dogs were trekking. Villumsen took the lead breaking trail through the snow, as he was especially adept at finding the route markers, most of which were almost totally covered in snow, with only a tip showing. Spotting the markers was more difficult because of the low level of light. Even at noon the sky was like evening twilight. They progressed only 5–7 kilometers (3–4.5 miles) per day for the first few days. By the seventh of October they had been sixteen days on the trail and had covered only 170 kilometers (106 miles) from

West Station—only slightly more than 40 percent of the distance. The temperatures ranged from -30° to -40° Celsius (-22° to -40° Fahrenheit) with little change from night to day because of the long night and low daytime sun. After three days of strenuous effort fighting the strong headwinds and deep snow, they stopped for a day of rest.

Their concern now was their diminished food supply. They had estimated an amount of food based on double their current rate of travel. Only by making better time could they last another two weeks on the trail. They talked of turning back to West Station and determined that the 230-kilometer (143-mile) marker was the point of no return. Beyond that point they would be better continuing to Eismitte even though westward travel would be easier by going with the wind. They knew that emergency food stores had been placed at intervals along the trail, but they would not consider breaking into them, because Georgi and Sorge would be depending on the emergency food supplies for their highly risky trip to West Station.

At the 170-kilometer (106-mile) point Wegener made the decision to turn back to West Station. He was sure they could not reach Eismitte before October 20, if at all. The next day the weather improved enough that they changed their minds and again headed east toward Eismitte. By this time the dog food (pemmican) was so low that they would have to go all the way to Eismitte even if they should meet Georgi and Sorge somewhere along the way. On October 13 they reached the halfway point and found the stored supplies of kerosene, gasoline, and the winter hut that the motor sleds had dropped off when they turned back on September 23.

They stopped for a day to rearrange loads, taking some of the supplies at the storage depot, and to kill some of the dogs. Now they were forced to revert to the practice of disposing of dogs as the need for them lessened during the trip. The meat from the dogs extended the time they could last on the trail.

On the sixteenth of October they passed the 230-kilometer (143-mile) point, their agreed point of no return, with weather holding suitable for travel. Loewe wrote that "our retreat now seemed to be cut off."[2] He was commenting on the lack of reserve supplies on the trail, except for the emergency stores that they had decided not to touch. Wegener expressed a hunch that Georgi and Sorge might stay at Eismitte after all. Loewe disagreed. He felt that Georgi and Sorge would leave on October 20 so everyone would know what they were doing. Wegener hoped that if Georgi and Sorge left Eismitte they could be persuaded to return there if they happened to meet them on the trail. If not, then Wegener was prepared to stay the winter at Eismitte himself along with Loewe. None of this could be resolved until they finally made contact with Georgi and Sorge either on the trail or at Eismitte.

On October 20, when Georgi and Sorge had said they would be leaving Eismitte, Wegener, Loewe, and Villumsen still had 111 kilometers (69 miles) to go to reach Eismitte. If Georgi and Sorge were traveling on foot, and if the weather continued to be good, they should meet them in about four days at the 335-kilometer (208-mile) point. By this time their load of supplies for Eismitte was reduced to one can of fuel, a tent, a bucket, and a lantern. Travel conditions improved as they continued east because the wind had created a hard crust strong enough to hold up dogs and men without breaking through.

The grind of Arctic travel was beginning to have its effect on the sled party. Men and dogs were weakening from the constant effort. Sleeping bags, clothes, and other gear were getting wet. Loewe had painful, though not yet serious, frostbite on his fingers and toes. Wegener was able to avoid frostbite altogether, perhaps due to his greater experience at taking care of himself in the Arctic.

In his book *Kabloona*, Gontran de Poncins describes the agony of facing Arctic winter winds. "I shut my eyes and pressed down my eyelids as if my purpose were to solder them together. When the wind is cutting your face into ribbons there is nothing else you can do. Certain parts of me—cheeks and chin, particularly—had begun to burn as if seared with a hot iron, and where the burning took place I felt the flesh suddenly harden. I was shriveling up. I tried to lower my head, to turn sidewise away from the wind, to roll up in a tense and miserable ball. I was ready to give up, and for a word I should have broken into sobs. My soul was shaken. Nature here was too strong, there was no resisting her."[3] Poncins wrote of the unrelenting lessons taught by life in the Arctic and the "inconceivable harshness" with which it inflicts its lessons. When a blizzard arose he found in an instant that one's mind becomes totally focused on survival.

On October 24 Wegener's party stopped at the 335-kilometer (208-mile) marker and waited for a day, thinking they might meet Georgi and Sorge there. The strong wind and bitter temperature of -40° Celsius (-40° Fahrenheit) gave them little desire to travel anyway. As dusk fell they walked toward the east keeping an eye out for the pair from Eismitte. Nothing appeared but the northern lights shimmering in hazy forms. Fritz Loewe reported that the beautiful lights moved Wegener to speak, as he rarely did, about his feelings. He spoke of the role of science in the advancement of mankind and of how his part in that process inspired all his actions.

He and Loewe also spoke at length about what to do if Georgi and Sorge did not come. They agreed that the Eismitte team must have changed their plans despite the letter. By this time turning back was out of the question. They were only 68 kilometers (42 miles) from their goal. But Villumsen had lost hope of reaching Eismitte and wanted to turn back. The problem of Greenlanders wanting to go home simply would not end. Eventually Villumsen agreed to go on. He had little choice as returning alone he would have had little chance of surviving.

As they began the last 68 kilometers (42 miles) of their trek, they were desperately low on dog food, with enough for only two and a half days at best. The temperature stayed at -50° Celsius (-58° Fahrenheit) all the time because the low sun angle and the short days allowed no heating during daytime. The moisture of their breath froze into ice crystals and new crystals from atmospheric moisture would form around their breath crystals, creating a trailing cloud of ice crystals for some distance behind them.

Cold fingers made trying to grasp anything painful and extremely slow. In this part of the Arctic it was the practice to use the fan-style dog harness with each dog attached directly to the sled, rather than the in-line harness used in the western parts of Arctic North America (figure 7.1). The fan-style harness allows dogs freedom to change positions back and forth within the team, thus becoming entangled. Disentangling harnesses was a constant effort for the Wegener party, but often it could be accomplished without stopping. Untangling dog harnesses was especially difficult because it had to be done with bare fingers, and teeth too if the lines became knotted. While camped the men would have to stop outside chores and take refuge in the tent periodically to warm their hands at the Primus stove.

Dogs also suffered from the cold, and needed more food as the temperatures fell. Often dogs would try to break into the tents to get food. Feeding the dogs was an arduous task, like every task in the Arctic winter. The dog food, pemmican or dog meat, was frozen so hard it had to be broken with an ax.

On October 27 Loewe noticed that his toes on both feet had no feeling. For several days Wegener gave lengthy massages to Loewe's feet, both morning

FIGURE 7.1
The fan harness system, which is used by the Greenland Inuits,
connects each dog directly to the sled. Tangled lines present a
frequent problem. This photo is not from the Wegener expedition.
(Photo by Fristrup, from Fristrup, 1966.)

and night, but circulation did not come back. Fortunately, the severe frostbite
did not prevent him from driving the sled and keeping up with the others, but
they decided that Loewe should stay at Eismitte over the winter.

On the evening of October 28 they used the last ration of dog food with
only 23 kilometers (14 miles) to go to reach Eismitte. They were determined
to get to Eismitte on the next day, but in the low light they couldn't find the
trail flags and were forced to camp only 5 kilometers (3 miles) from their destination. They had jettisoned the loads farther back to increase their speed, but
still could not finish that day. On the morning of the thirtieth they used the
last of the kerosene to heat some black pudding for breakfast. Later that
morning Georgi heard the crunch of snow outside the ice cave and shouted,
"There they are."[4] He ran down the corridor and out into the open air in his
underclothes. There stood Rasmus Villumsen with his dogs and sled. Rasmus
laughed and pointed toward Wegener and Loewe coming behind him.

When they reached Eismitte fog was on the ground and temperature was
at -52° Celsius (-62° Fahrenheit). They had taken forty days to make a trip

that could be made in fourteen days in good weather. Wegener was in excellent condition, high spirits, and proud of having reached Eismitte under such difficult weather. Villumsen also was in excellent condition. Loewe was exhausted, had badly frostbitten toes, and his heels had turned blue. He had marched for the past four days with frozen toes. Loewe went into the Eismitte ice cave, crawled into his sleeping bag and stayed there for several months. When the group entered the ice cave they were nearly overcome by its warmth of -5° Celsius (23° Fahrenheit). Wegener was so pleased with the Eismitte situation, he kept repeating, "You are comfortable here! You are comfortable here."[5] He repeatedly told Georgi and Sorge how very glad he was that they had decided to stay at Eismitte for the winter.

When Wegener told the story of their trip, he described Loewe's frostbite, the dogs' hunger and fatigue, and the temperatures in the range of -50° to -54° Celsius (-58° to -65° Fahrenheit). He said he had never heard of a more difficult trip in Greenland. Georgi wrote, "How valuable radio communication would have been. Now, unhappily, three men have incurred the gravest danger."[6]

Wegener spent several hours writing in his diary and making notes on the meteorological observations that Georgi and Sorge had taken. They spent most of October 31 eating and drinking coffee in the ice cave, and discussing the situation for the men remaining at Eismitte. Certainly Loewe could not travel and would have to stay for the winter. In addition to causing some crowding in the ice cave, his unexpected stay would mean a more rapid consumption of the food supplies. Provisions considered barely adequate for two men would now have to be split three ways. Also, they decided to give 135 kilograms (300 pounds) of food provisions and a can of kerosene to Wegener and Villumsen for their return trip to West Station. These decisions were based on careful calculations of supplies and needs for everyone concerned.

On November 1 they celebrated Wegener's fiftieth birthday with some photographs (figure 7.2), fruit, and chocolate saved for special occasions. Then Wegener and Villumsen set off westward with two lightly loaded sleds and seventeen dogs. The wind was at their backs, and the temperature was slightly warmer at -39° Celsius (-38° Fahrenheit). It was an ideal day, and they were confident that travel would be easier going west with the wind. After a short distance they disappeared into a light ground fog, and the Eismitte crew turned to the tasks that would be facing them for the winter. Georgi felt

FIGURE 7.2
Alfred Wegener and
Rasmus Villumsen on
the day of departure
from Eismitte,
November 1, 1930,
Wegener's fiftieth
birthday. (Used with
permission of the Alfred
Wegener Institute for
Polar and Marine
Research, Bremerhaven,
Germany.)

deeply moved at Wegener's departure because of the danger the sled party would be facing. As he watched them leave he was anxious about the safety of his friend and longtime mentor, knowing that it would be May before he knew if Wegener and Villumsen had reached West Station. Georgi wrote, "I need not gloss over the fact that, after the two sleds disappeared in the fog, I retired to the barometer room to compose myself."[7]

## Wegener and Villumsen Alone on the Trail

As ground fog closed around the two dogsleds, the Eismitte weather tower was soon out of sight behind Wegener and Villumsen. Thoughts of farewells and birthday best wishes were quickly replaced by the urgency of the journey facing them. Two hundred fifty miles stretched between Wegener and the base station on the west coast of Greenland. Though the weather was now calm, November 1 was not a time to expect good weather. The eastward trip to Eismitte had lasted so long their food supplies now did not allow for any delays. If the winter storms resumed, the return trip could be a disaster.

Alfred Wegener and Rasmus Villumsen had a plan to conserve food for themselves and the dogs. They would begin with two sleds to carry as much food as possible. As individual dogs weakened from the meager rations, they would be killed and become food for the remaining dogs. When the number of dogs became too few, Wegener would abandon his sled. He would then continue on skis while Villumsen drove the other sled. Wegener had no doubts about his skiing ability or about keeping up with the dogsled. He was in the prime of health and proud of his physical condition. His main wish was for the dog food to last long enough that the distance on skis would not be too great—perhaps only halfway.

Wegener and Villumsen knew they could travel fifteen to twenty miles a day if they did not encounter any storms. By traveling westward they would keep the wind at their backs and have a gradual descent all the way to the West Station. On the trip to Eismitte the Arctic's most wretched headwinds had slowed their progress to a crawl. They had every reason to feel optimistic as they started westward, but haste was essential if they were to outlast their food. The prospect of fierce winter storms and -50° Celsius (–60° Fahrenheit) temperatures were additional reasons for haste.

Weather was not to be in their favor. Lower temperatures with driving winds would have drained the dogs' energy, and the added stress of low temperatures would take its toll. The dogs would have faltered one by one. Each morning Wegener must have discovered to his dismay that more dogs had died. The dead dogs would have been cut up for dog food. Wegener would have weighed the need for food against the need to lighten the load. He chose to jettison a box of pemmican to lighten the load and conserve energy. At this point they had moved only seventy-three miles from Eismitte, and their progress was slowing.

They abandoned one of the sleds after another nineteen miles and harnessed the seven remaining dogs to Villumsen's sled. Wegener put on his skis. They had now progressed only ninety-two miles from Eismitte—far short of the hoped for halfway point to the west coast. Wegener would have tried not to slow their progress further just because he was now on skis, so they probably attempted to maintain the same rate of travel as before. At this exhausting pace, he probably began to lag behind by the end of the day. When the few hours of dim daylight ended, Villumsen would have stopped first, begun pitching the tent, and a short time later Wegener would have caught up. It would be sure death to continue after dark without visible trail markers.

Perhaps it took as much as two weeks to travel 212 kilometers (132 miles)—65 kilometers (40 miles) of it since Wegener donned the skis. Their progress must have slowed to about 6 miles per day. When Wegener reached the camp that day, he must have been exhausted. He probably ate a quick meal of pemmican and crawled into the tent to sleep. The next morning when Villumsen got up and began preparing for travel, he must have thought it strange that Wegener was not yet out of his sleeping bag. When he went to check, Wegener was dead.

Rasmus Villumsen must have been distraught when he discovered Wegener's body. He had been with Wegener on a previous expedition and no doubt mourned the loss of a respected friend. He would also have realized that he was alone on the immense ice cap and he would have had a very real fear of getting lost in its vast, featureless spaces. Villumsen dug a shallow grave in the ice and buried Wegener's body wrapped in two sleeping bag covers. Villumsen covered the grave with snow and marked it by sticking the skis and ski poles upright over the grave. He collected Wegener's personal items including important notebooks and disappeared into the whiteness.

Villumsen apparently traveled thirteen miles before dark and pitched camp. In the morning Villumsen must have doggedly tried to continue in the ongoing storm, but it was futile. Rather than risk travel in these conditions he made camp again after going less than a mile westward. He must have realized he would have to wait for the storm to end, and he seems to have stayed in this same camp for three days.

When the wind slowed, Villumsen attempted the trail once again. He had little choice but to push on or face starvation. The wind must have kicked up a blinding ground blizzard of snow, and the trail markers became ever more difficult to find as drifting snow covered them. At some point Villumsen would have come to the realization that it had been too long since he last saw a trail marker, and he would have turned back to find the previous one. Swirling snow with no landmarks or sun for guidance soon would leave Villumsen completely disoriented. In such conditions it was almost impossible to travel in a straight line, but Rasmus would have felt certain that he was going straight back to his last camp, even though the snow had obliterated his tracks.

Finding no evidence of his previous camp or any sign of a trail marker, he would have known he was hopelessly lost. What to do? He could stop and camp until the storm ended. That might take days, perhaps outlasting the remaining food. He could keep trying to find the trail with the risk of moving away from the trail completely. As with most people in this dilemma, Villumsen probably chose to move on. Movement gives a sense of doing something. With luck, he could have happened on a trail marker, but he probably never found one. By the end of that day he must have known the end was near for him. It is likely that Villumsen packed up and continued moving through the storm. He was never seen again.[8]

# *8*

## *Winter at East Station and West Station,*

## *1930–1931*

*I want to go south, where there is no autumn, where the cold*

*doesn't crouch over one like a snow-leopard waiting to pounce. The*

*heart of the north is dead, and the fingers of cold are corpse fingers.*

——D. H. LAWRENCE

### *East Station*

Walther Kopp and two other men sailed from Copenhagen on July 10, 1930, and arrived at the mouth of Scoresby Sound on the nineteenth. Because the sea ice stays so much longer on the east coast of Greenland, no ship can enter Scoresby Sound before the middle of summer. Kopp had experience with high-altitude atmospheric measurements at the Aeronautical Observatory in Lindenburg, Germany. Hermann Peters was trained as a biologist, and Arnold Ernsting was an engineering student.

At the mouth of the sound they learned that pack ice would prevent them from reaching the planned location for the East Station. They unloaded all their gear near a small Inuit settlement and pitched tents to wait for the ice to melt. On July 28 their ship *Gertrud Rask* departed to make its rounds of the villages of Greenland. Kopp and his companions were left on shore with piles of equipment and provisions meant to last for a year. Their equipment included thirty-seven miles of steel wire, fifty bottles of hydrogen, six gasoline engines with fuel, and a winch—all for launching tethered balloons for weather observations. They also had a radio, batteries, and a prefabricated wooden hut.

While waiting for the ice to break up, they set up a meteorological station to begin collecting data. This was a hedge against the possibility that the ice might not open enough to reach their intended destination some 140 kilometers (85 miles) up Scoresby Sound. Using a small motorboat, they made regular trips to test the condition of the ice. During their wait, they lost some of their supplies when their two small boats broke loose during a storm and drifted away with cargo they had loaded in preparation for departure. In another instance, one of the boats hit an iceberg and sank. In each case the boats were recovered, but the saltwater ruined seventy days' worth of provisions.

One difference in life on the east coast was the presence of musk oxen and geese, which the expedition members could shoot to supplement their food provisions. However, in midwinter when they most needed extra food, the musk oxen had disappeared. Finally on September 1 the ice in the sound cleared, and the next day the three departed in the boats to establish East Station. On the fifth of September, halfway to their destination, they established a storage depot of provisions to be available for their spring trip back to the settlement at the mouth of the sound. Because the boat back to Denmark would pick them up in early summer before the pack ice melted in the sound, they would be returning to the settlement by dogsled rather than the small boats. The storage depot would be necessary for that trip in the spring.

On September 18 they reached the designated location for East Station. They immediately began erecting the hut as signs of winter were well upon them. On the thirtieth they finished the hut in time for the first snow. The next day they established the weather station, and set up equipment for raising tethered balloons. They rigged an antenna for the radio and managed to establish contact with the Danish at the Scoresby Sound station. Through the Danish operator, they could hear information about the West Station. In this way they heard that Wegener was on a sled trip to Eismitte. On November 28 they contacted West Station by radio directly for the first time, and soon they had arranged twice weekly messages. They learned that Wegener had not returned from Eismitte.

Despite losing some provisions, they had enough food and fuel for the entire winter with ground temperatures continually around -40° Celsius (-40° Fahrenheit). They had a special Christmas feast of eel omelet, stewed musk ox with cabbage, potatoes, and plum pudding. On New Year's day they listened to celebrations on their radio from countries as far away as Japan.

On March 9 Peters had an attack of acute appendicitis. When they notified the Danish by radio, a relief party set out to bring Peters back to the settlement, but they had to turn back because of severe storms. Fortunately, Peter's pain subsided in a few days, and no medical emergency developed.

In March the men at East Station began to run short of food, and the musk oxen were not around. They estimated that the food supply would last until the end of April. Kerosene for heating and cooking was getting low, and they began to think they might have to abandon East Station early. A message from the settlement near the mouth of Scoresby Sound advised them to return no later than the end of May because the river ice would melt, making swollen rivers difficult to cross. Furthermore, after that time the sea ice would soften enough to become unsafe for foot travel.

On May 10 they left East Station with three sleds and three dogs for the trip over land and sea ice back to the mouth of Scoresby Sound. In the face of storms, snowdrifts, snow blindness, and low food supplies, they reached the depot of provisions on the fourth day. Now they had sufficient food for ten days, including enough for the dogs, which had been without food for at least a day. With food plentiful again, they had a feast of frozen sausages and herring. On the seventeenth they reached the settlement, where they waited four weeks for the arrival of the ship to return home.

## West Station

A s the fourth sled trip with Wegener, Loewe, and thirteen Greenlanders left for Eismitte on September 22, the West Station crew was finishing the hectic work of moving tons of equipment and provisions up to the station. The need for the ponies was almost finished, and they would soon be slaughtered for winter meat. The pony handlers, Vigfus and Jon, had already returned home to Iceland. The summer melting of the ice had ended, and the portable winter hut could now be built. The expedition's motorboat *Krabbe* was taken to Umanak where it could be beached for the winter to prevent it from being crushed when the fjords froze.

The Greenlanders returning from the fourth party brought a letter to Weiken from Wegener dated September 28 at the 62-kilometer (39-miles) site. Despair with only a shred of hope can be detected in Wegener's letter.

Dear Weiken,

My worst fears have been realized. Not only have the motor sleds not got beyond 200 kilometers [125 miles], but our sledge party too has broken down owing to the unfavorable weather. Of the 12 Greenlanders who remained after Ole left us, eight are returning today. We had great trouble persuading the other four to stay with us, and whether we shall manage to reach the Central Station with them remains to be seen. This morning we have a temperature of -28° Celsius [-19° Fahrenheit], drifting snow and a headwind; lovely weather! Please get back everything of ours from those who are returning. As before, they are to get 4 kroner a day; I have promised 6 kroner a day to those who are going on with us. We shall now try to take the necessary kerosene to the Central Station, and other than that only odds and ends. Whether Sorge and Georgi will be able to remain there or will come back with us remains to be seen. If it is at all possible, I should like the station to be maintained through the winter. We cannot take the hut, but they were willing to do without it. The whole business is a big catastrophe and there is no use in concealing the fact. It is now a matter of life and death. I will not ask you to do anything to ensure our safety on the return journey, for there are plenty of depots. The only help you could render us would be the psychological one of sending a party out to meet us; but in October that would again involve considerable risk to the relief party. I do not consider Sorge's plan of setting out on 20th October with man-pulled sleds feasible; they would not get through but would be frozen to death on the way. We shall do what we can and we need not yet give up all hope of things going well. But good traveling conditions seem to be definitely over now. Even the journey here was very strenuous, and what lies ahead of us is certainly not a pleasure trip.

Best wishes to all and may we all meet again safe and sound with some satisfactory achievements to look back on!

Your Alfred Wegener[1]

In a second letter written at the same time, Wegener gave a list of all the material and supplies they were leaving at his present position and said he expected to reach Eismitte on October 14 and be back to West Station by October 25. This proved to be an unrealistic outlook. Instead of taking a large sled party to Eismitte with all the necessary supplies and equipment, Wegener was down to only the most essential item, kerosene for heating. Wegener was concerned that Eismitte would have to be abandoned without additional kerosene. Weiken and others at the West Station were gravely concerned on hearing this news. Doubts were cast on the success of Eismitte and on the safety of Wegener and others in the fourth party. Both motor sleds were disabled: one at 51 kilometers (32 miles), and the other at 41 kilometers (26 miles) from West Station. They could not be retrieved and repaired until spring, maybe as late as May, when travel was again possible.

Movement in the West Station was difficult enough. No one had anticipated the great depth of snow in the Scheideck area. Much of the snow was blown in by the east wind from the surface of the ice cap. Ponies could no longer move at all, and dogs moved only with difficulty. At this time the ponies were slaughtered. Flanks and shoulders were stored for the men to eat; the rest was for the dogs.

Wegener sent another letter with the last three Greenlanders who refused to continue. It was sent from the 150-kilometer (93-mile) depot. The gloomy outlook seems to have dissipated somewhat in this letter dated October 6.

Dear Weiken,

The soft deep fresh snow has reduced our rate of marching very much: October 1, 15 kilometers [9.3 miles]; October 2, nothing; October 3, 6 kilometers [3.7 miles]; October 4, 14 kilometers [8.7 miles]; October 5, 11 kilometers [7 miles]. This has thrown our program in a heap again. We are sending three Greenlanders home. I had promised each of them a watch from the expedition if they carried on to the 200 kilometer [125 miles] depot. Seeing that *we* are now sending them back, please give them the watches they have been promised and see that a watch is put aside for Rasmus, who is going on with us.

We are going on from here with three sledges, which later will be reduced to two, and hope in this way to be able to reach Georgi and Sorge (although with practically no supplies for them) either at Eismitte or on their way back to the west station. This plan has the following advantages:

1. The winter station at Eismitte will be maintained even if only as

a meteorological station and little else; for Loewe and I have decided to winter there in the event of Georgi and Sorge being unwilling to do so. True, we have only 1.3 liters of kerosene a day, so conditions will be very primitive, but in our opinion, not too risky.

2. Georgi and Sorge will have one or two dog teams and a Greenlander to accompany them and look out for the various depots which are available. In our opinion, this is the only way of ensuring their safe return; otherwise they would perhaps get stuck in the region of heaviest snowfall and lose their lives.

3. We shall be relieved of the distracting and unbearable uncertainty whether Georgi and Sorge have had the sense to stay where they were, or tried to return and perished on the way. I am thinking of press telegrams and the people at home.

You now know definitely to expect three men to return, as there are not enough provisions for five at Eismitte.

Please send out a small relief party with dog sledges, say two dog teams and two members of the expedition. The relief party should make its base at the 62 kilometer [39 mile] depot and await the return of ourselves or our comrades there. It would be very desirable if food depots could be laid farther on, as the whole route beyond 62 kilometers is only sparsely provided with them. The relief party should leave Scheideck about the 10th of November, and must be prepared to wait at the 62 kilometer depot until the 1st of December. During the time of waiting perhaps a measurement of the thickness of the ice or some other scientific work could be carried out. We have rations for a marching rate of 12 kilometers [7.5 miles] per day, a rate which we are assuming for the return journey also in view of the shortness of the days. This would bring us back to the 62 kilometer depot on the 21st of November, but we hope to arrive earlier.

We are all well—no frostbites so far—and hopeful of success. Do not let yourself or your comrades be tempted to run risks in pursuing your scientific work. Please get the relief party to improve the route marking or stick new flags in the area.

Greetings to all.
Alfred Wegener[2]

The measurement of ice thickness that Wegener referred to was one of the most innovative aspects of the expedition. The thickness of the Greenland ice cap was measured by seismic sounding techniques for the first time. A dynamite charge was exploded, and a portable seismograph recorded the travel time of the shock wave as it moved down to bedrock and was reflected back to the surface. By knowing the velocity of seismic waves in ice, they could calculate the depth to the rock.

Weiken, who had been left in charge at West Station during Wegener's absence, could now see that the fourth sled party was not achieving its mission. Clearly Wegener was continuing to Eismitte because he thought Georgi and Sorge were planning to abandon it on October 20. Wegener felt the trip that Georgi and Sorge said they would make on foot pulling small sleds was impossible in the terrible weather. He was concerned for their safety first, but a second motive was that someone should man Eismitte. If not Georgi and Sorge, then he would do it himself.

On November 10 Weiken and Kraus decided to go out to meet Wegener at the 62-kilometer (39-mile) marker. They took the Greenlanders Mathius Simeonsen and Johan Villumsen, brother of Rasmus. For this trip they also took a radio to keep in touch with West Station in case Wegener should show up there without seeing them on the trail. The first day a storm hit and they barely made the short distance across the crevasses to a place they referred to as "start point." The dogs simply could not make it in the strong wind and drifting snow. They waited four days in a tent for the storm to pass.

By November 16 they had moved as far as the 35-kilometer (22-mile) marker, and at noon saw the sun on the horizon for the last time. It was the beginning of the Arctic night. The next day they got as far as the first abandoned motor sled, the Eisbär, at 41 kilometers (25 miles); the storm had intensified so much that they stayed there three days. On the twentieth they passed only two flags, then saw no more—the snow had drifted over them. They crisscrossed the supposed route and eventually found a snow cairn at 50 kilometers (31 miles) from their starting position, and the second abandoned motor sled, the Schneespatz, was just a short distance beyond, almost buried in drifts. Soon they lost the markers again and could not find the 55-kilometer (34-mile) marker where some gasoline had been stored. They were hoping to mix some gasoline and kerosene to make flares for Wegener and Loewe to see as they approached. Eventually they gave up and pitched tents for the night.

On November 21 they found the trail but soon lost it again. They needed to reach the 62-kilometer (39-mile) marker that day to be sure they did not miss Wegener. The chances of getting there that day were bleak as they could see barely a hundred meters ahead.

In the dim twilight of the Arctic day Weiken became separated from the others. Being separated and lost in the Arctic is a life-threatening situation. He thought of going back alone to West Station without a tent or a stove. He traversed back and forth across where he thought the trail should be and scanned along the sharp crests of the drifts, looking for tracks where the other sleds might have passed. After what must have seemed a long time he saw two forms coming toward him. Kraus and the Greenlanders had come back to look for him, and they just happened to intersect as Weiken was again traversing at right angles to the trail looking for tracks. Kraus had found the marker where Wegener asked them to wait. After twelve days (with only six of travel) they had arrived at the rendezvous point. This was the day that Wegener had estimated he would arrive at that point. The cached supplies there had not been touched, so they assumed that Wegener had not yet arrived.

For the next two days Weiken and Johan Villumsen laid out closely spaced flags on a line extending about 10 kilometers (6 miles) both north and south of the trail, so that Wegener would be sure to intersect them even if he had deviated from the direct line of the trail. Kraus and Mathius built an igloo using blocks cut from the ice and moved out of the tent into the warmer igloo. Filling all the cracks with snow kept the wind out much more than a tent could. By running a Primus heater they had the temperature in the igloo above freezing near the roof and holes began to appear from melting. They adjusted the heater to create a workable inside temperature, which was -8° Celsius (18° Fahrenheit) on the sleeping ledges. They covered the inner door with reindeer skin and covered the outer opening to the vestibule with a provision box to better block drifting snow and invading dogs.

Everyday as twilight turned to dark night, about 3 P.M., they lit a kerosene flare for three hours as a signal for Wegener to see should he pass near. By 6 P.M. anyone traveling would have stopped to camp for the night, so they extinguished the flare. They considered going ahead during the day on the chance of meeting Wegener earlier, but they realized that would only increase the chance of missing him altogether, considering their own difficulty staying on the trail line.

On one clear day (no sun, but good visibility) Kraus and Mathius took some provisions to a more easterly marker on the chance that Wegener had to camp less than one day's travel from them. They carried only flags and provisions (food, dog food, and kerosene) to leave at the marker. To lighten their own load, they took no tent, sleeping bags, or provisions for themselves. Fortunately, there was no storm, and they returned easily, though after the full darkness of night. Weiken was concerned for them, as the wind had increased causing snow to drift again. He had lit the regular flare and had walked east to light another flare when he saw them in the distance. Kraus and Mathius had made a round-trip of 36 kilometers (22 miles) in the twilight without getting lost. Kraus had a good sense of direction with stars visible, but good luck was also with him.

Kraus unpacked the radio and did the necessary warming of all the parts over the heater. All switches, contacts, coils, and batteries had to be warmed before the radio could operate. Every day the message from West Station was the same: Wegener had not arrived there. It was highly improbable that Wegener would miss them, as he had a sled odometer and a sextant for determining his position, and he was an expert in navigation. Also, cairns and flags were more visible to anyone traveling west due to the east wind keeping them swept clear. They speculated constantly about Wegener's situation. Knowing he was in top physical condition and that he had a lot of Arctic experience, they concluded that he should survive the trip from Eismitte. Perhaps he had elected to stay at Eismitte for the winter, but they knew this was doubtful considering the limited supplies there. They waited at the rendezvous site until December 5, five days past the time Wegener had requested. They were able to return to West Station in only one day, as opposed to twelve days coming out.

The men of the West Station were anxious and depressed when Weiken and Kraus returned with no word of Wegener. Nonetheless, most chose the optimistic explanation that Wegener had remained at Eismitte. They knew a rescue trip was out of the question and that Wegener himself was strongly opposed to winter relief trips that would endanger the larger group. The men sought to focus their minds on the hard work at hand. They each knew the overall objectives of the expedition and how to make their individual contributions. So the project proceeded without Wegener. They entertained themselves with music and conversation in their leisure time.

But with Wegener's absence, they missed the chance to learn from him about science in Arctic regions, where none of them had experience. Because they did not know exactly what his own scientific work was to have been during the winter, they could not gather any data that he personally might have wanted. They proceeded with their own tasks in meteorology, glaciology, seismology, and geodesy. They wanted to assume that Wegener was safe, but they could never feel certain that this was so.

Apparently Wegener's party had no knowledge of a British expedition, led by H. G. Watkins, going on at the same time about 483 kilometers (300 miles) south of Eismitte. The British created a meteorological station manned by A. Courtauld, who wintered alone on the ice cap. The British expedition included two airplanes for pilots to gain experience flying in the Arctic and to take aerial photos. They built their station at the highest point of the potential east-west air route across Greenland. Their intent was to survey the eastern mountains and ice sheet elevations for a future air route. The urge to be first with an air route prompted the British group to keep this expedition secret as much as possible. A proposed air route would start from London, go to the Faeroes Islands, to Iceland, then across Greenland about one-quarter the distance from the south end, and across Hudson Bay to Canada. Some airlines eventually flew that exact route.

The British had many of the same supply problems as Wegener's group, so only one person was able to man the winter station. Dogsled drivers to the British station were inexperienced and had trouble making headway bringing supplies. Because of fuel leakage, Courtauld discovered he had enough kerosene for light and cooking but not for heating. Though they planned to have a radio, it didn't make it to the station. Courtauld's cave was so covered with snow by March that he couldn't get out and had to sit in the cave until rescued. During that time he made no meteorological measurements. In April he ran out of kerosene and had to sit in the dark until mid-April. When a relief party attempted to find his cave late in March, they had to return to the west coast because the storms prevented their search, even though they knew they were actually near the site. Then the leader of the group, Watkins, attempted to fly in for a search but engine trouble made him turn back. They returned by dogsled on April 18 and spotted the vent. They called down the vent, got a response from Courtauld, and dug him out. Since they had no supplies to continue the station, it was abandoned.

# 9

## Winter at Eismitte

*If we survive dangers it steels our courage more than anything else.*

—B. G. NIEBUHR

Georgi and Sorge were alone at Eismitte from September 13 until We-gener arrived with Loewe and Rasmus Villumsen on October 30. The weather was relatively mild and a tent was adequate while they excavated an ice cave. By the time they began moving into the ice cave on October 5, morning temperatures were down to -37° Celsius (-35° Fahrenheit), and the improved shelter was most welcome.

On the day before they moved into the ice cave, they saw an Arctic fox scavenging among their provision boxes. This was a rare sight in the center of the Greenland ice sheet, as no regular wildlife inhabitants live there. They assumed the fox had followed a trail of food scraps left by sled parties and their dogs. Arctic foxes are known to travel great distances following polar bears or wolves to eat leftovers from their kills. This fox was snow white—they turn brown in summer—and appeared to be in good health even though it was about 400 kilometers (250 miles) from its natural food sources on the coast. The little animal showed almost no fear of the men, and they were able to photograph the fox from as close as 10 meters (33 feet). Later it would come out of its den dug in the ice, and come as close as 2 meters (6 feet) to take pieces of whale meat they offered. Other explorers have found Arctic foxes so tame in behavior that they often became a nuisance around campsites. During the day the fox spent time near the warm air from the vent of the ice cave. After a few days the fox disappeared and was not seen again. They never knew if something happened to it or if it suddenly headed for the west coast again. The only other sighting of wildlife at Eismitte was the brief appearance of a snow bunting. One day it appeared, fluttered around their camp several times, and flew away.

Georgi and Sorge continued to expand the ice cave, finding it easier to dig than they had expected. Digging was done by cutting blocks of ice from the walls and floor of the room. The upper layer of the glacier is composed of snow compacted to firn, which had the consistency of damp, cooked rice that has cooled and congealed. The firn was easy to cut and shape with a knife, saw, or spade. They cut blocks from the interior, carried them outside, and built a wall around the entry as a barrier to blowing snow. Eventually they had enough ice blocks to build an observation tower for mounting the theodolite to be used in tracking the weather balloons. After each trip carrying an ice block from the deepening room to the outside, they had to rest; they were at an elevation of 3,000 meters (9,840 feet). A thick layer of firn was left in place to form a roof for the cave, and they built supports from below to help prevent collapse during settling. On the inside of the cave they had raised platforms for sleeping and storage. The observation tower gradually grew to a height of 3 meters (10 feet), and became known as the "castle of the Holy Grail" (figures 9.1 and 9.2). The dug-out rooms became the "dungeon." Georgi photographed various stages of development of the building operation at Eismitte but could not develop the film for lack of a darkroom. At night the temperatures were much too cold for film processing.

They built a primary room 3 meters by 5 meters (10 feet by 16 feet) for sleeping, cooking, and writing. A slightly smaller room was built for storing provisions and supplies. A third room near the entry was used for inflating the meteorological balloons, and the smallest room was for hanging a mercury barometer. A central corridor connected these rooms. At the end of the corridor was the beginning of a shaft that Sorge used to observe the seasonal and density variation of firn as it compacted into ice.

Daylight filtered through the roof giving the cave an eerie blue light. The men felt they were inside a white marble crypt. When they needed water they cut a piece of firn from the wall and melted it on the stove. If they needed a place to hang clothing, cooking pots, or instruments on the wall, they stuck a wire or wooden peg into the firn. They created an air passage for ventilation by pushing a ski pole through the roof and controlled the air movement by a piece of metal that could swing over the opening. Appropriate ventilation was a tricky problem. If the vent was closed too far, fumes from the cooking stove began to overpower them. If they opened it too wide, a gale of cold air swept through the cave making it impossible to heat. During storms and high winds

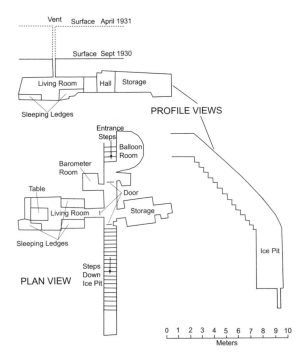

FIGURE 9.1
Floor and elevation views of Eismitte give an idea of the effort expended in establishing the Central Station. (Adapted from E. Wegener, ed., *Alfred Wegeners letzte Grönlandfahrt,* 1932.)

FIGURE 9.2
This surface view of Eismitte shows the instrument box and the observation tower, called the castle. (Used with permission of the Alfred Wegener Institute for Polar and Marine Research, Bremerhaven, Germany.)

they would close the vent entirely, as the doors leaked enough to keep air moving through the cave.

Heating the cave was another skill Georgi and Sorge had to acquire. They had to keep the temperature below freezing at the ceiling level to prevent it from melting and dripping water on everything below. The result was that the temperature at table-top level was -5° Celsius (23° Fahrenheit) and on the floor, -15° Celsius (5° Fahrenheit). Compared to outside temperatures in the -50s or -60s Celsius (-58 to -85 Fahrenheit), this was relatively cozy. However, even when they were working they kept sleeping bags wrapped around themselves against the cold of the cave.

When he made the decision not to leave Eismitte, Georgi knew the fourth sled party was probably already on the way. He regretted this, but had no way to stop them from coming. The failure of the motor sleds became the focus of blame as seen in these notes from his diary: "We all counted so absolutely on the motor sleds. When one thinks of the unbelievable work which those two vehicles caused; the transfer to another ship at Holsteinborg, on the edge of the ice at Uvkusigsat, and the transport difficulties on the glacier, one cannot help feeling sorry for Wegener in the disappointment of such great expectations. How much more advanced everything else might be if we had not all 'backed the wrong horse!' Now, of course, there is nothing more to be done, and the last dog sledge trip, which is bound to be difficult, is the penalty for our common deception."[1]

Each day as they went out for weather observations they looked to the west hoping to see the approach of the sleds. After a few weeks at Eismitte they found they could get by on less fuel than estimated. They could cut their heating time from ten or twelve hours per day to only one or two hours in the afternoon. They accomplished this by spending more time in their sleeping bags between tasks. They recalculated their needs against their available supplies and concluded that they could make it until late May, by which time a relief party should have arrived. They estimated that they could get by on 1.3 liters (1.4 quarts) of fuel per day, allowing for some losses. Wegener's original estimate (with the winter hut for shelter) allowed 4.7 to 6.2 liters (5 to 6.5 quarts) daily.

All these considerations played a part in their decision not to abandon Eismitte on October 20. They were also motivated by their dedication to the objectives of the whole expedition. This was to be the first major Greenland expedition with more than a year's meteorological data from west coast, center, and east coast stations. The thrill of being the first was too much to miss. The risk of trekking so far on foot in the winter also played a big part in the decision to stay. Georgi's notes explain his reasoning: "Meanwhile the position has changed in several ways. We have dug ourselves into the snow, and thus are protected against all cold and assaults of the weather, and by exercising a certain degree of economy we can hold out until spring with our remaining eight cans of petroleum. We have ample food supplies. On the other hand, in the horrible weather which has prevailed for several weeks, the journey of 400 kilometers (250 miles) back to Scheideck with hand sledges involves considerable danger, certainly much greater danger than remaining here at the station."[2]

Another factor was loyalty to Wegener himself. Wegener had a personality and an enthusiasm for his work that inspired his colleagues. He had given them enough responsibility and decision-making authority that each had a strong personal stake in the success of the project as a whole. Their only regret in making the decision to stay at Eismitte was that they did not have a radio so they could communicate the change of plans to Wegener at the West Station. They hoped the fourth sled party would not try to come at all. The temperatures had dropped; the winds had increased; and they knew it would be a difficult and hazardous trip. Though Wegener eventually made it through, he brought nothing to help the fuel situation, and even had to take some of Eismitte's food for his return trip. In addition, Loewe remained at Eismitte to share the limited food supply.

Eismitte had only 1.3 liters of kerosene per day until spring. Wegener had meant for them to have five times that much, but had not been able to get it there. Also, while en route to Eismitte, Wegener had cached the instruments, Christmas packages, a phonograph, and the radio, leaving a bleak winter ahead for the inhabitants of Eismitte. There would be hours of idleness, weeks of loneliness, no mail or communication from the outside, and eventually depression and despair.

## Three Alone

By the ninth of November Loewe's toes had wasted away and become shapeless. Sinews were sticking up like ridges in the decomposing flesh. Loewe spoke constantly of the need to amputate, and they all agreed that the toes would have to come off. They had no book on frostbite or medical procedures, no surgical instruments. All these items lay cached somewhere on the trail between them and West Station with all the other goods. They had a tiny supply of cotton and bandage material. The only cutting tools they had were some metal shears and a pocket knife that Georgi began sharpening to a razor edge. They had a bit of iodine for disinfectant, but no way of deadening pain. That night there was little sleep for the two who would be surgeons, nor for Loewe, the patient.

The next morning they lit the stove and put it near Loewe's sleeping ledge so he would not have to expose his bare leg to the freezing temperatures of the room. They melted ice and heated the water for washing. They tried applying snow to Loewe's feet to serve as a local anesthetic, lacking anything better. When that had no significant effect in reducing pain they realized they had to proceed anyway. Georgi cut away the flesh around the base of the toes of the right foot with his pocket knife, then nipped the bones of the second and fifth toes with the metal shears. Sorge steadied Loewe's leg and held the light. Loewe endured this procedure without anesthetics and afterward was talkative and in good spirits. Georgi wrote regarding Loewe's bravery, "Not for nothing does he possess the Iron Cross, First Class," referring to Loewe's award for courage in World War I.[3]

In a few days the left foot had further deteriorated, and they knew those toes had to come off too. The operation was repeated with equal success. They were anxious that perhaps they had not removed enough decayed tissue, and the mortification might continue to spread. Georgi and Sorge both felt some responsibility for Loewe's misfortune, as they had been the reason for Wegener and Loewe's dangerous winter trip.

The rest of the winter consisted of a routine of scientific observations. Georgi got out of his sleeping bag at 7:35 A.M. and went out for measurements of temperature and wind. If a blizzard was blowing there was a real chance of getting lost even in the short distance between the cave door and the weather station. Visibility reduced to nothing in the darkness and swirling snow caused disorientation. In fifteen minutes he was back and announced the weather as a wakeup call to all in the room. Georgi's temperature announcement provided a sense of expectancy as they waited to see what would be the record low.

One morning Sorge went out to take the morning observations in a raging snowstorm. Georgi could hear from within the ice cave that Sorge was moving farther away from the entrance of the ice cave rather than coming closer. Georgi quickly got out of bed, dressed, and took a light to signal the direction to the door. Luckily, Sorge had wandered in a big circle and ended near the entrance of the ice cave. It was dark and visibility was so bad that the ice tower many feet high could not be seen even two steps away. Later that day they marked the track to the weather station with poles, rope and flags, and squirted coffee dregs on the wall of the tower to make it more visible in snow.

Georgi was the breakfast cook. He started with a hard rye-crisp bread, *Knäckebrot*, which he toasted and buttered and served to Sorge and Loewe while they were still in sleeping bags. Oat porridge was Georgi's specialty. He served porridge with apricots, porridge with prunes, porridge with lemon drops, porridge with chocolate, porridge with coffee, porridge with leftover bread, porridge with onions, porridge with dried bananas, porridge with plum juice, porridge with orange peel, watery porridge, and thick porridge. One consistent ingredient was reindeer hair that shed from the sleeping bags and clothing and wafted about the room. They ate their porridge with a spoon in one hand and tweezers in the other.

After meals Georgi and Sorge spent most of their time repairing and adjusting instruments and recording data in notebooks. When the cold began to chill them too much, they had to climb back into the sleeping bags to warm up. Georgi was the most experienced at instrument repair and modification. He made a great breakthrough by modifying the clocks driving the recording instruments so they would operate reliably in cold temperatures. Poor clock operation below -45° Celsius (-49° Fahrenheit) had been a common problem with all previous Arctic expeditions. By adding an extra spring, Georgi was able to get uninterrupted records throughout the winter, including the record low.

Developing photographic plates was another sensitive task in low temperatures. First snow had to be melted to create developing, fixing, and washing baths. If the wet plates froze in the low room temperatures while drying, the ice crystals destroyed the emulsion. So the plates had to be put in a drying oven. If the glass plates got too warm, the emulsion would come off. These procedures used quite a bit of extra fuel. Georgi lamented the loss of many valuable photographs while he was learning the correct procedure. Figures 9.3 and 9.4 give some idea of the working conditions inside the ice cave.

FIGURE 9.3
Johannes Georgi and Fritz Loewe inside Eismitte, Christmas 1930.
(Used with permission of the Alfred Wegener Institute for Polar
and Marine Research, Bremerhaven, Germany.)

FIGURE 9.4
Johannes Georgi at work inside Eismitte. (Used with permission
of the Alfred Wegener Institute for Polar and Marine Research,
Bremerhaven, Germany.)

For evening meals Sorge was the primary cook. They usually ate canned meat and vegetables in a stew, or meat rolls of corned beef or pemmican. Contents of cans were always frozen rock hard and had to be thawed. Meal preparations usually included enough food for two days to minimize cooking effort. The leftover meal, which froze as soon as it was set aside, required heating but no additional preparation.

Special occasions, such as birthdays or anniversaries of almost anything, called for some variation in the food. The favorite was frozen whale meat and an apple or orange. They commented that the whale meat tasted like venison, and the frozen apples and oranges sounded like billiard balls when they were hit together. Fruit was the most prized treat of all, and they had brought enough to have some every Sunday. Like all food, fruit had to be thawed before it could be cut, peeled, or eaten.

Loewe's feet were beginning to heal, although he still had a lot of pain. Georgi had to perform some minor follow-up cutting to remove additional bone and rotting flesh. In the beginning they changed his bandages daily, but as healing began they changed them at three-day intervals. Bandage material was so scarce it had to be cleaned and reused—which meant using fuel to melt and heat wash water. Though the amputations healed very slowly, the wounds never became infected. They attributed this good fortune to an extremely low germ count in the subfreezing room temperatures. Likewise, no one caught a cold through the entire winter. When Loewe finally made his first attempt at walking, after three months in bed, he could scarcely stand alone. With daily effort he began to regain strength in his legs.

One thing they had in abundance was lice. Eismitte was free of lice until Loewe arrived. His close contact with the Greenlanders during the trip was the source of his infestation. One evening Loewe picked 370 lice from his body, clothing, and sleeping bag. Fortunately the lice could not survive a trip from Loewe's bed to either Georgi or Sorge because of freezing temperatures. This helped keep the critters isolated. Every few days they took Loewe's sleeping bag outside and shook the lice to the ground where they froze instantly. This had to be repeated frequently because of lice on Loewe's body and eggs remaining in the sleeping bag, but it kept the numbers down to a bearable level. Loewe used a hot iron on his clothing to kill lice.

Late in December Georgi broke a tooth, which soon became infected. The entire left lower jaw was festered and swollen. The tonsils became swollen and his lower lip was completely numb. Georgi heated a pair of pliers red hot with his soldering flame in order to bend them enough to reach in and remove the root. Despite two hours of probing the swollen gums with pliers and a screwdriver, he was unable to get a grip with his makeshift dental instruments. Soon the swelling spread over the entire left side of his face. The tooth problem eventually cured itself after many days of pain and swelling.

During this same period, Loewe experienced some abdominal pain that they all feared might be appendicitis. Again they were lucky, and the pain disappeared. For these three men, isolated, without medical knowledge or supplies, and with no chance for rescue, luck seems to have been their main source of help.

One of their pastimes was reading from a small library with selections from science, literature, and history. Loewe liked to read aloud poems by Goethe or Schiller, and the others enjoyed his skill at reading with feeling and emotion. They agreed their isolation from the abundance of cultural stimulation in Europe had heightened their sensitivity to art and literature. They regretted that they had no way to play music in their cave, though they would occasionally sing together. One evening Sorge read aloud from a book about Scott's disastrous experience and death in Antarctica. Hearing this story affected Georgi emotionally, and gave him a bad premonition about Wegener's situation. He had to ask Sorge to stop reading.

From November 21 to January 23 they had no sun. About noon during those months the sky would lighten slightly, but the stars were still visible. The one advantage of continuous darkness was the frequent splendid display of northern lights.

On December 15 they celebrated Georgi's forty-second birthday, and everyone looked forward to some treats. Candy, fruit, and candles made the day special. They also celebrated Sorge's thirty-second and Loewe's thirty-seventh birthdays at Eismitte. They celebrated Christmas with a display of artificial flowers, boxes of candy, and fruit. They each had saved some family letters, which they read for the first time as part of their Christmas observation. This period marked the beginning of a growing loneliness.

When the sun began to reappear in January, warming did not accompany it. Rather temperatures continued to get colder, and the record low occurred at the equinox on March 21: -65° Celsius (-85° Fahrenheit). The sunshine, however, was a great tonic for their spirits. They began to get outside more and even took some short ski excursions. In his book *Kabloona*, Poncins described the same positive feelings when sunlight returned to the environment: "Perspective was here again. The scene had a foreground, a middleground, a background."[4]

A surprising and unnerving phenomenon of the ice cap in winter is strong tremors caused by sudden settling and compaction of the firn under the increasing weight of accumulating snow. A thundering sound followed by a windlike roaring accompanied the ice tremors, and the ceiling of the ice cave settled slightly. The needles of the recording weather instruments left a trace of the shock like a seismograph. When the first of these tremors occurred, they feared that the roof would collapse under its heavy accumulation of snow estimated at 20,000 kilograms (22 tons). The weight alone made the roof settle about 6 centimeters (2.5 inches) per month, and the shock of an ice tremor might cause it to fall on them. The roof had dropped downward to the top of the bookcase. Georgi immediately built an ice pillar in the middle of the room. Though this put their minds at ease, they may still have been at risk. They guessed that the loud noise that accompanied the ice quake was caused by air suddenly forced out of the looser layers of firn. In their study shaft, uncompacted firn was observed down to 6 meters (20 feet). Below that the firn was much denser and harder.

De Poncins, in *Kabloona*, described the effects of continuous darkness of Arctic winter as a great weight. "Bit by bit this silence that had soothed me and laid to rest my frayed nerves, began to seem to me a weight. The horizon was closing in round me. The prison, once so radiantly peaceful, was now unveiling its true face. . . . Now, in the dead of polar winter, the line I hesitated to cross was drawn in a radius five feet around the stove."[5] He also wrote of the immense effort required to venture outside, even for the most menial tasks, while ground blizzards blew a wall of snow around him. "Emptying the slop bucket was a trial. Whether your destination lay ten yards or ten miles off made no difference in your preparations; and the disproportion between those preparations and the aim of your errand was grotesque."[6]

Barry Lopez, in *Arctic Dreams*, describes the oppression of continual darkness and persistent wind. "The darkness, worse than the cold, shuts out the far view and drives one deep into his clothing and shelter. This can be a time of mental depression and occasional sudden violence. Oral traditions of the Inuit have many nightmare images associated with the winter months."[7]

Winter at Eismitte was a continual battle against the ever-present cold. The subfreezing temperatures combined with darkness diminished their effective living environment down to the dimensions of the room. As loneliness and homesickness grew, their spirits swung from high to low, then anger followed by apathy. Georgi's notes express his despair. "We feel ourselves abandoned here. Wintering here in the middle of Greenland is indeed no trifle. And it is unfortunate that we should have to do it now under such unfavorable circumstances! We are very depressed, especially at not receiving the whole of our summer mail and our Christmas parcels."[8]

For Georgi, Sorge, and Loewe, the saving grace was their passion for the work they were doing and their warm comradeship. These kept them motivated to make the experiment a success. Sorge, optimistic, energetic, and enterprising kept the group entertained with his readings of Schiller's poetry. Loewe, who felt constant pain for months, tended to be more melancholy, but he made witty remarks even when complaining of the pain. He entertained the group with his extensive knowledge of expeditions to the Antarctic.

Amid the long hours of darkness, beautiful moments existed as well. Georgi described stepping out into the mooonlit Arctic night:

> An hour ago I took a walk outside. The full moon in the eastern sky, burnished silver, seeming to smile scornfully at the northern lights which cast several broad curves from east to south, but whose pale light was faint compared to the moonshine.
>
> As I walked along in -55° Celsius [-63° Fahrenheit] the surface of the snow groaned and cracked with every step, and now and again the sharp sound of a little snowquake caused by the walking—the subsidence of the uppermost strata of the snow—was audible far around. My breath formed thick clouds, and far away on the western horizon, a broad cloud of vapor, hung the trail of mist that issued from our chimney. The whole effect was indescribable. Nature here is so completely alone, and pays no attention to us tiny intruders.[9]

# 10 ❖

## The Search for Wegener and Villumsen

*A man's dying is more the survivor's affair than his own.*

—THOMAS MANN

Winter storms eventually became less intense and less frequent. It was time to take provisions to the men at Eismitte and find answers to the big questions. Where were Wegener, Loewe, and Villumsen? How had they and the scientists at Eismitte fared through the winter? Men at the West Station were prepared for any possibility. Maybe Wegener, Loewe, and Rasmus died trying to reach Eismitte. Maybe Georgi and Sorge died attempting to return to West Station. Maybe they were all safe at Eismitte. In case anyone was still at Eismitte, they would certainly need provisions for the coming summer work.

In preparation for the trip to Eismitte, Jülg and Weiken had to make a preliminary trip on April 3 to cache stores of dog food at snow cairns along the way. They could not possibly carry enough dog food and other provisions in a single trip. Even though the weather had improved, they needed to stop frequently and wait for blizzards to pass. Last year's trail marker flags were nearly all covered with snow. They went as far as the 120-kilometer (75-mile) marker and left large quantities of both dog food and human provisions for the use of the relief sled party.

The spring trip to Eismitte began on April 23, 1931, with seven sleds and eighty-one dogs. More and more they felt certain that all the men had stayed at Eismitte. On his way to Eismitte with the dogsleds, Weiken observed some signs of Wegener's last trip. At 190 kilometers (118 miles) inland from West Station, Weiken found Wegener's skis standing up in the ice about 3 meters (10 feet) apart with a broken ski pole halfway between. He had dug down about one meter, but found only an empty box. At this time he had no expectation of finding a grave, so he stopped digging. They also found where Wegener had abandoned a sled at the 255-kilometer (158-mile) marker—the point where Wegener began skiing. Weiken found a box of pemmican dog food at the 285-kilometer (177-mile) marker—only 115 kilometers (71 miles) from Eismitte. Weiken may have suspected the worst but chose to ignore the clues. He seems to have interpreted the evidence on the trail to mean that Wegener, Loewe, and Villumsen had made it to Eismitte and stayed there for the winter.

On April 25 Kelbl, Kraus, and some Greenlanders went out to the abandoned motor sleds. They had to fit new pistons on the spot and take the sleds back to West Station for a thorough overhaul. By the end of April both motor sleds were ready to travel. Then they had to wait for a storm to pass, and by this time the dogsled party had an eight-day start on them. One motor sled was equipped with a small radio.

By noon of May 1 both motor sleds started eastward up the rise of the ice cap. Kelbl drove Schneespatz and Kraus drove the Eisbär, each with a Greenlander in the backseat. Every little rise taxed the capability of the sleds, and deep drifts required skillful maneuvering to avoid getting stuck. The worst drifts were always on the upward slopes, and in one of these they both got stuck at the same time. The driver had to get out, push, and then nimbly jump back in the driver's seat, being careful to avoid the unshielded propeller in the rear.

Bad weather reduced the visibility so they could not see the route markers, which were 500 meters (1,640 feet) apart. In low visibility that was far enough that the next flag could not be seen, and they might veer off the trail. Although they could have guided by compass, they camped until the storm passed. On the next day, May 5, they began to make good progress.

When they got to the 200-kilometer (125-mile) site they discovered that the rear runner supports on the sleds were so bent they were ready to break. They had to move loads forward and drive more slowly over bumps. They refueled there and carried some more gasoline with them. The glacier's surface became flatter as they continued eastward, and the crust on the snow made travel easier.

The motor sleds caught up with the dogsled party, and they camped together for one night. The next morning they took off in a race to reach Eismitte. Of course the motor sleds outdistanced the dogsleds even though the dogsleds left earlier. They were thinking Wegener would be quite pleased when he saw how well the motor sleds were performing.

As the motor sleds approached Eismitte on May 7, Sorge heard them and ran out of the ice cave to greet them. Kraus and Sorge ran to each other with heartfelt hugs as they simultaneously asked, "Where's Wegener?" The silence answered the question for them both. Johan Villumsen suddenly realized his brother was not here and that he must be dead. He was distraught, but no one knew enough Inuit to console him. As Loewe appeared from the ice cave, they realized that Wegener and Rasmus were the only ones missing. Weiken now saw the significance of the abandoned equipment, supplies, and Wegener's ski poles in the snow.

That evening the dogsled party arrived. The silence at the camp told its own tale. They ran up to the tents of the motor sled party shouting, "What's the matter?" There was no answer. Then Loewe limped outside and said, "Wegener and Rasmus left for the west on the 1 of November, so they are dead."[1] They all gathered in the ice cave and consoled themselves with conversations about Wegener and Rasmus. They spoke of how much Wegener meant to them, and how much he had cared for them.

Kraus set up the small radio he had brought and on the first try contacted Godhavn 600 kilometers (370 miles) away. He tapped the sad news on his sending key, and Godhavn relayed it back to Germany. The group at Eismitte spent hours discussing what had happened and what to do next. They decided that the motor sleds would return directly, taking Loewe with them as he was unable to endure a long trip. The dogsled group would conduct a search for the bodies during their return trip. Georgi would stay at Eismitte through June and July to obtain a complete meteorological record for central Greenland. Sorge would leave Eismitte and join the Weiken dogsled party in the search. Everyone agreed that they should carry on with the objectives of the expedition. Figure 10.1 shows part of the spring relief party assembled at Eismitte.

They knew that Wegener had left Eismitte with two sleds and seventeen dogs, and that he had hoped to reach the halfway station with both sleds before abandoning one sled and continuing on skis. They estimated that Wegener would be trying to cover 25 to 30 kilometers (15 to 20 miles) per day in this fashion for the last half of the distance back to West Station.

FIGURE 10.1
In the spring of 1931 the motor sleds arrived at Eismitte to bring
relief supplies. Part of the relief party is shown with the motor
sled Eisbär. (Used with permission of the Alfred Wegener Institute
for Polar and Marine Research, Bremerhaven, Germany.)

Now that they knew for certain that Wegener was dead, they would go
back and search more thoroughly for traces of his last trip and try to find his
body. They also wanted to try to find out what happened to Rasmus.
Holzapfel, Kraus, Loewe, and two Greenlanders returned to West Station with
the motor sleds in record time—only sixteen hours' driving time with one
night on the trail. With the snow surface hardened, and the winds subdued,
motor sleds could at last function as intended.

In three days Weiken, Sorge, and five Greenlanders reached the site where
Wegener's skis protruded from the snow. This time they dug deeper than before. First they began to find reindeer hairs in the snow suggesting a parka or
sleeping bag nearby. Then they found a reindeer skin and Wegener's fur clothing, which was spread over the top of his sleeping bag. Weiken wrote: "We
found Wegener's body sewn up in two sleeping bag covers. It was lying on a
sleeping bag and a reindeer skin three quarters of a meter [2.5 ft] below the

snow surface of November, 1930. Wegener's eyes were open, and the expression on his face was calm and peaceful, almost smiling. His face was rather pale, but looked younger than before. There were small frost bites on the nose and hands, such as are usual on journeys like his."[2]

Wegener was fully dressed with extra kamiks on his feet and dog skin trousers over cloth trousers. On his upper body were a shirt, skiing tunic, waistcoat, woolen jacket, thick sweater, and woolen wind jacket. Also he had on a wool helmet and cap. The clothing was in perfect order and free from drifted snow. The boots were still soft and not iced up. He lay on his sleeping bag, not in it. His pockets still had some items in them, but his pipe, tobacco, diary, a bag of personal items, and his fur gloves were missing. The missing diary no doubt had interesting and useful information about his trip out to Eismitte, and the journey back to the point where he died. Now they would never know the details of Wegener's last few days.

All evidence indicated that Wegener did not die on the march, but in his tent. Probably he died from heart failure due to overexertion trying to keep up with the dog team. Else Wegener wrote in 1960 that, in her opinion, Alfred's heart had been weakened by the cross-country ordeal during the expedition of 1912–1913 and by the stress of war wounds. Although she did not mention smoking, one cannot help noticing that many of his pictures show him with a cigarette or pipe. Nothing in the record indicates anything less than ideal physical condition for Alfred. On that final journey he probably had endured much more cardiac stress than could most healthy people.

Rasmus also must have been in good condition up to the time of Wegener's death. Weiken and Sorge were deeply touched by the care with which Rasmus had buried Wegener and by the way he arranged and marked the grave. Rasmus had thoughtfully taken Wegener's bag of personal items and his diary, intending to bring them to the West Station. He must also have taken Wegener's gloves, which were better than his own.

Weiken and Sorge carefully sewed Wegener's body back into the bags and returned it to the ice. Above the grave they built a large vault of firn blocks. One of the Greenlanders planted a small cross made of Wegener's broken ski pole. Then they tied a black flag to the skis. They made these markers to be sure they could find the site again. Later they returned and erected a large cross made of iron rods about 6 meters (20 feet) high (figure 10.2).

Continuing westward they were able to find two sites where Rasmus had camped. At 170 kilometers (105 miles) they found reindeer hairs and scraps of pemmican. That they could find these items after six months of wind and snow is nearly miraculous. Only 1 kilometer (0.6 miles) farther west Rasmus appeared to have stayed for several days, as they found remains of several meals and a hatchet. At 155 kilometers (96 miles) from West Station they found no trace of a camp but found evidence that dogs had spent considerable time there. None of the food caches between this site and West Station had been touched.

At the end of May, Weiken, Jülg, Sorge, and three Greenlanders again set out eastward over the ice cap. Jülg and Sorge were to continue their geodetic survey begun last summer. Weiken and the others went to look for Rasmus's body. During early June Weiken and the Greenlanders made daily traverses north and south of the trail. They drove back and forth in arcs of 30 to 50 kilometers (18 to 30 miles) with empty dogsleds. They started at the last known site where Rasmus camped and gradually worked westward, searching the horizon with binoculars. They planned a search pattern that would not omit any area from their view. Low undulations in the surface, with long

FIGURE 10.2
Wegener's grave site after the cross was erected. (Used with permission of the Alfred Wegener Institute for Polar and Marine Research, Bremerhaven, Germany.)

troughs between ridges, limited the view. To be sure they had seen into all the troughs, they made numerous side trips out from their traverses. In eight days they saw no trace of Rasmus, and by now their dog food was low.

Weiken sent the three Greenlanders and their dogs back to the West Station and continued the search by himself for another eight days. Weiken was bothered by many illusions in the snow depending on the direction he looked relative to the sun. With his back to the sun all the snow drifts looked like cairns. Looking into the sun the surface looked like a mountain landscape. The view at right angles to the sun was most deceptive as every snow drift had a white side and a dark blue side. In these cases every object looked like a man-made object—a cairn, a sled, even a tent. Weiken's imagination began to play more tricks on him. Sometimes he felt he was surrounded by cairns, tents, or sleds either at rest or moving across the landscape. Sometimes he drove the dogs at full speed toward a snow drift feature that looked like it had been a camp site. He then dug all around, pushing a long brass rod into the snow repeatedly; eventually he was forced to admit that it was no different from any other drift.

Weiken had now used up almost all the dog food, so he headed for West Station. The search for Rasmus had failed, and he was depressed. In addition to the pain of losing Rasmus, Weiken felt defeated by the loss of Wegener's important notes and diary. While at Eismitte, Wegener spoke of many interesting things he had observed and entered in the diary. Entries made on his last segment of the trip would undoubtedly have revealed information about the tragedy.

What had become of the twenty-two-year-old Rasmus? He could have become disoriented and traveled in the wrong direction until food and fuel gave out. He could have reached the western part of the ice cap with its large crevasses, and fallen through the snow cover that often hides them. By the end of winter all traces had been covered and further search was futile. Weiken's search might have passed very near Villumsen's body hidden under a drift.

Georgi, staying at Eismitte, would not know what the search party found until the sleds came for him in August. Until then, he was alone. He speculated that Wegener had died of carbon monoxide poisoning. He supposed that snow might have drifted over the tent, sealing off sufficient air for the stove, causing it to burn badly and fill the tent with carbon monoxide. Wegener had mentioned such situations. Georgi wondered if there might have been a mishap with loss of food and fuel. These ideas weighed on him heavily.

Already he had been criticized for his decision to stay at Eismitte in October after announcing he and Sorge would return to West Station on foot. Though Wegener was pleased that they stayed, Loewe and others felt they should not have changed their minds once they had given warning of abandoning Eismitte. In November, Sorge and Georgi had both suggested that one of them should return with Wegener and Rasmus, but Wegener could see no justification for taking them away from important work at Eismitte, given the good physical condition he and Rasmus were in.

Georgi was unable to shake a sense of guilt about his decision. Alone at Eismitte, Georgi wrote, "I see ghosts, cannot bear the dark, and have fastened the door for the first time in nine months. I am doing very little work."[3] Clearly he was severely shaken by his long isolation and the tragic news with its incriminating insinuations.

Georgi wrote that his fatigue and loneliness during the long isolation at Eismitte used "the last fiber of one's being; one cannot just take part or stand aside. Twelve months of inland ice do wear a man out."[4] He later began hearing noises and imagining someone was approaching: "What a lot of noises one hears here. Today I was quite sure I heard the buzzing of the motor sleds and dashed up—but there was nothing there. It was the wind, or a kind of buzzing in my ears. How often I have thought I heard something in the night. Certainly the dog, Bella, makes all kinds of noise turning the rubbish heap upside down, but mostly it is pure imagination."[5] Bella was one of the sled dogs that had been left behind because she was disabled. She apparently survived by scavenging in the trash pile for food scraps and feces carried out from the ice cave.

By mid-July Georgi began to plan what he should do if a relief party failed to come for him soon. They were already past the expected time. He set a date of July 31, modified a sled with skis, and revived the old plan for walking back to West Station. He hooked the dog, Bella, to the sled, but she seemed to have forgotten what to do. After many trials the dog began to pull. Georgi also built a sail, because he would be traveling with the east wind. Since he did not have a tent, he planned to sleep during the day when the sun was shining and travel at night.

This careful planning was not needed, however, as both motor sleds arrived on July 24. They spent several days packing, while continuing to collect data, and abandoned Eismitte forever on August 7, 1931.

# *11* ❖

## *Searching for Reasons*

*Things are where things are, and as fate has willed.*

*So shall they be fulfilled.*

— AESCHYLUS

Alfred Wegener's brother, Kurt, took the job of project director when Alfred died. This contingency arrangement had been made in the contract for the expedition. Upon news of Alfred's death, Kurt prepared for a leave of absence from his job in Hamburg and arrived in Greenland in July 1931. The project proceeded through the summer; Georgi returned from Eismitte in August; then both the West Station and the East Station closed, and all personnel returned to Germany. Kurt had the task of coordinating each member's contribution with the final report of the expedition.

Some expedition members felt that the final work would have gone more smoothly if one of them had filled in as director. Karl Weiken had directed the West Station during Wegener's absence. When Fritz Loewe returned from Eismitte, he became interim director until Kurt Wegener arrived. They may have felt reluctant to transfer authority to an outsider when so much inner group loyalty existed. This issue faded quickly as Kurt's final report on the expedition was praised by all.

Johannes Georgi wrote his views on the subject of succession of authority on expeditions:

In the event of the leader's removal his successor should be taken from the ranks of the expedition itself. Every plan worked out in a study at home is, in its execution, modified by all possible circumstances, and not only by external conditions, but quite often by psychological factors. These things are common property of all the members of the expedition; they are discussed for months on end, but hardly ever completely set down in writing. How can a leader coming fresh from home to the expedition know of these imponderables, so important to the result, if not decisive? How can he know the views of members of the expedition themselves after their performances during the expedition, or of the efficiency of native auxiliaries, or of the points at which political factors may come into play?

It should be a regular rule, and one which need involve no ill feeling, that the leader should, right at the start, appoint his second-in-command in the presence of all members of the expedition and clothe him with full authority in case any unexpected disaster should overtake himself.[1]

Georgi wrote a letter to the expedition's sponsors, the Emergency Committee for German Research, asking them to appoint a committee of polar explorers to inquire into the case of Wegener's last journey. He explained that some at West Station said the fatal journey had not been intended but had been caused by those at Eismitte making demands in excess of the original program. The fact was that Wegener had promised Georgi and Sorge that if the motor sleds failed, a fourth sled trip or more would be made if necessary in the autumn of 1930. Georgi felt that many of the expedition members probably never knew this.

In his book *Mid-Ice* Georgi wrote of his admiration for Wegener: "His example and his personality inspired his comrades to stake the whole of their strength, and all their personality for a great common objective; . . . all of them, while always obliged to recognize their great leader's superiority, felt ennobled and magnified by his confidence; and no quarrel ever disturbed our collaboration—these things will keep alive the memory of Alfred Wegener' expedition and its great leader, so long as men go out to northward and southward into the realms of ice."[2]

The Emergency Committee for German Research held the hearing that

Georgi requested. Various members of the expedition testified on their understanding of events leading to Wegener's death. A few criticized Georgi for not returning to West Station when he said he would. Georgi responded that the critics did not understand the peculiar conditions of Arctic exploration and that Wegener himself understood that external conditions control men's actions and often force them to change plans.

Georgi also heard criticism that things other than fuel had been taken to Eismitte, forcing Wegener to make a special unplanned trip to bring more fuel. Georgi demonstrated that Wegener had approved everything that had been transported in each trip to Eismitte. There had been some misunderstanding among some of the members over how much had been intended for transport to Eismitte, since they had not been included in the preparations for the trip.

Georgi later wrote some interesting comments on the issue of decision-making on a major expedition, in which members of the expedition are widely separated:

> Arctic expeditions generally extend over such a wide area that
> weather, snow, and ice conditions can no longer be judged from the
> spot where the leader is. This means that the actions of each separate
> party must to a large degree be dictated by the local circumstances, so
> that the military system, with unrestricted power of command in the
> hands of the leader, is not suitable. On the contrary, it must be the aim
> of the leader of an expedition so to impress his plan and his wishes
> upon each separate party that, in an emergency, it can work inde-
> pendently, utilizing local conditions in the best possible manner in the
> interest of the general plan. Alfred Wegener always promoted inde-
> pendent action on the part of separate groups.[3]

Both Wegener and Georgi knew the situation at Eismitte was precarious from the beginning. Both men had relied on the success of the motor sleds that failed in the autumn of 1930. If it had been known sooner that the motor sleds were not adequate, Wegener could have arranged additional and larger dogsled parties in late summer and early autumn. The announcement by Georgi and Sorge that they would return to West Station if not sufficiently supplied by October 20, would not have deterred Wegener from going out there himself to manage Eismitte. Data from Eismitte was such a high priority to him, and so essential to the whole expedition, that he would have spared no effort to get it.

The project had half a million German Marks supporting it, and Wegener would do his best not to let it fail. Wegener once expressed the thought that the cause must succeed whatever might happen. He believed men must be willing to make sacrifices for the goals of the project and that strong commitment for the common good would bind the group together. He referred to this as a religion for expeditions that ensured there would be no regrets. This expedition certainly tested his creed.

If Georgi and Sorge had left Eismitte on October 20, Wegener still would have made the trip out there, but he would not have had to return to West Station. He would have stayed along with Loewe and Rasmus until spring. Hence, he would not have made his fatal trip back to West Station. Whether Georgi and Sorge could have survived the trip on foot and pulling a sled is another question. This trip also carried a high risk. It probably would have taken at least thirty days (double the normal dogsled trip) assuming they were able to carry enough food and fuel for that time. That would have taken them into the terrible November storms that Wegener faced. Although the Emergency Committee for German Research cleared Georgi, and all members of the expedition shook hands in a spirit of unity, Georgi suffered from doubts for many years to come.

Criticism of Georgi had validity only in hindsight. With the information available to everyone at the time, Georgi and Sorge made a logical decision that would have worked if winter had not begun earlier than expected that year. In his final report Kurt Wegener made no criticism of Georgi and reported favorably on his scientific contributions. The committee concluded that no guilt fell on any member of the expedition and praised the hard work and loyalty of the entire party. The committee considered the matter closed, but Georgi continued to hear of negative comments from former members of the expedition. Friedrich Schmidt-Ott, president of the committee, met with Georgi, Sorge, and other members of the expedition—along with Else Wegener. The president concluded that he had succeeded in making peace among them. Perhaps he brought peace, but not healing.

In 1960, thirty years after Alfred's death, Else wrote a biography of Wegener in which she mentions her concern that others might have contributed to her husband's death. As a mourning widow, she was probably biased, and her opinion should not hold more weight than that of expedition members. Nevertheless, her comments reopened painful wounds for Johannes Georgi. Referring to the last trip to Eismitte that Wegener made, Else wrote: "Since on this trip serious decisions had to be expected as leader of the expedition, he [Wegener] felt responsible to take part in it himself. They were supposed to bring the most needed items, the shelter and fuel. Georgi himself took over the selection of transported goods for Eismitte. Why on earth did he not have them bring the most important things first to get them through the winter? It was now the end of September. There was no longer nice summer weather, and a smooth surface without wind. Every day became shorter and would bring bad weather. It was clear to me what this meant. Now the most difficult trip fell to Alfred."[4]

Else recognized, however, that luck had been against the expedition from the start. She referred at one point to the long delay in April while waiting for the ice to thaw in the fjord before they could unload. "The six weeks of waiting had shifted all calculations. Six weeks earlier a fourth dog sledge trip would have easily brought the winter hut and the fuel! Six weeks earlier in Kamarujuk all transport work could have been done in a fraction of the time, without pressure and overexertion. Six weeks earlier!"[5]

From a scientific viewpoint, the expedition was a success. The compiled data provided meteorological insight to the region that influences weather for much of northern Europe. The meteorological and glacial data obtained have provided benchmarks with which weather and ice thickness data are still compared today. Wegener's expedition was the first to use seismic methods to measure the thickness of an ice cap. The expedition members also pioneered the use of weather balloons in the Arctic. They were the first to collect a full year of meteorological data from the center of the ice cap. The British expedition during the same year had a data gap and did not provide a complete year of data.

Johannes Georgi, leader of Central Station, returned to Germany in autumn of 1931 and continued his job as head of the Instruments Office with the German Sea Monitoring Center in Hamburg. He became co-organizer of the Second International Polar Year in 1932–1933, fifty years after the First

Polar Year in 1882–1883. He was the first scientist to report on large cirrus clouds forming from condensation trails of jet airplanes. Georgi kept in touch with Fritz Loewe and frequently mentioned their experiences in Greenland. He died in Hamburg at the age of eighty-three in May 1972.

In 1932 Ernst Sorge returned to Greenland with the Franck expedition and continued glacial studies. In 1935 he headed an expedition to Spitzbergen. He died in April 1946 at the age of forty-seven.

Fritz Loewe, a Jew, emigrated from the Third Reich with his family soon after returning to Germany—ironic, considering he had won the Iron Cross for outstanding bravery in World War I. He went first to the Scott Polar Institute in Cambridge, England, and eventually he started the Meteorological Institute at the University of Melbourne, Australia. Loewe joined a French expedition to Antarctica where he did his second overwintering on an ice cap. After retirement from the University of Melbourne, Loewe traveled to various places in North America and Europe as a guest lecturer. In 1971 he gave the keynote address at the annual conference of the German Association for Polar Research—his subject: Alfred Wegener and modern polar research. He died in 1974.

Else Wegener published three books after Wegener's death. In 1932 she edited *Alfred Wegeners Letzte Grönlandfahrt*, a compilation of accounts from several members of Wegener's last Greenland expedition and notes from Alfred's diaries. This was translated into English in 1939 as *Greenland Journey*. In 1955 she wrote a biography of her father (*Wladimir Köppen, Ein Gelehrtenleben für die Meteorologie*). Then in 1960 she published a book of Alfred Wegener's diaries, letters, and her own memories titled *Alfred Wegener: Tagebücher, Briefe, Erinnerungen*. In 1992, at the age of one hundred, she was made an honorary member of the German Polar Research Association. She died later that year.[6]

In 1942 a squadron of World War II U.S. fighters and bombers crash-landed on the surface of the Greenland ice cap. The crew members were all rescued by dogsled but the airplanes were left behind. Fifty years later one of the fighter planes was dug out by hydraulic mining techniques using warm water. In those fifty years the plane had been covered by successive years of snow changing to ice to a depth of 82 meters (268 feet). The downward compression of the ice had flattened the entire plane like a smashed can. The accumulation plus the downward movement of ice in fifty years amounts to an average 1.6 meters (5.2 feet) per year. One can extrapolate what must have happened to Wegener's grave site in the more than seventy years since he was buried. It must now be about 117 meters (383 feet) below the surface.

Over the years Wegener's grave site will have moved deeper and westward as the glacier moves slowly from a central north-south axis down and out toward the coasts. Though global warming will likely accelerate the movement and decrease the thickness of the ice, in a few centuries the grave could be near the west coast.

# 12 ❖

## Remembering Wegener

*Let us not underrate the value of a fact; it will one day*

*flower in a truth.*

—HENRY DAVID THOREAU

*A new scientific truth does not triumph by convincing*

*its opponents and making them see the light, but rather because*

*its opponents eventually die, and a new generation grows up*

*that is familiar with it.*

—MAX PLANCK

Serious discussion of continental drift came to a standstill after the American Association of Petroleum Geologists' conference in 1926. A few geologists, such as the two South Africans Alexander du Toit and Lester King, continued to support the theory of continental drift, but to the great majority, it was a dead issue until the 1960s.

Real scientific support for lateral continental mobility came somewhat inadvertently after World War II, when oceanographic research with vastly improved instruments received an enormous boost from national governments, particularly the U.S. Navy. This research was not motivated by anything Wegener had written, and testing the concept of lateral movement was not an initial objective.

During World War II the U.S. Navy became interested in all aspects of the ocean, and afterward seafloor mapping became a priority for submarine tracking. This impetus turned into a massive, worldwide data collection effort. Initially, the purpose was only to produce worldwide seafloor topographic maps. Soon maps of magnetic variations, seismic activity, and seafloor topography provided an array of maps that had never before existed. A door was opened, giving researchers a new view of the ocean. Continental maps were now augmented by maps of the ocean floor.

Research before the twentieth century was conducted almost entirely by individuals. Great movements forward in science were associated with names like Galileo, Darwin, Newton, or Curie. In most cases these researchers were working alone, using their own resources, or they were financed by a patron. After World War II research was strongly identified with national defense and was generously financed by governments. The United States Office of Naval Research opened soon after the war expressly for the support of oceanographic research. The navy in particular wanted to know more about the behavior of sound in the water (sonar), geophysical surveys (magnetic and seismic), physical and chemical characteristics of the water, ocean currents, sediments, and structures on the seafloor.

Research by interdisciplinary teams became the norm rather than lone scientists working in isolation. Not only was there interaction among members of a team but also there were frequent meetings between teams and exchange of personnel from one institution to another. Collaboration was the style of the day.

Grant money allowed the collection of significant amounts of data, opening a productive era of inductive research driven by data gathered from the oceans. Interpretation of seismic data from a global network of seismograph stations intended to monitor underground nuclear explosions provided insight into the internal structure of the earth. The global seismic network provided valuable information on natural earthquakes and the locations and depths of active epicenters around the earth. For the first time, these seismic data provided a complete picture of patterns of global earthquake zones. Magnetic data, meant to help in submarine detection, provided key maps showing patterns of reversals of Earth's magnetic polarity. These magnetic patterns led to the discovery of rifting and spreading of the seafloor along midoceanic ridges. Gravity field measurements, beginning in the 1920s, revealed anomalous gravity values in oceanic trenches of the western Pacific. Interpretation of gravity anomalies suggested that lighter oceanic crust was moving downward into the denser underlying mantle, producing lower gravity values over the trench. This was one of the first bits of evidence that the crust was not static but was moving, at least near the trenches.

In addition to the flow of money for research, instrument quality improved, providing greater sensitivity and resolution, which allowed detection of smaller features at greater depths. Improved detection proved vital to discovery of magnetic patterns on the ocean floor that would have been missed with previous instruments.

Another feature of the changed style of research after World War II was the sudden influx of large numbers of U.S. scientists emerging from universities, having studied under the G.I. Bill. This investment in education gave great returns in scientific research after the war. The increased number of institutions having significant research funds and numerous graduate students led to many advances in all fields.

Oceanographic research in the study of Earth's history was concentrated in a few institutions like Columbia University's Lamont (later Lamont-Dougherty) Geological Observatory, University of California's Scripps Institute of Oceanography, Princeton University, and Cambridge University in England. Most of the funding for the American institutions came from the Office of Naval Research and the National Science Foundation. Much of the funding was used for outfitting ships with instruments for collecting geophysical data around the world.

These changes set the stage for a better understanding of ocean floors than ever before. The objective was mapping. The result was maps that set off a revolution in earth science. No one began the various projects with the notion of testing a hypothesis about continental drift, but the outcome was irrefutable evidence for movement of huge crustal plates over the surface of the earth. The concept changed from that of continents plowing though oceanic crustal layers, as suggested by Wegener, to one of continents and oceanic plates moving together as units—plate tectonics.[1]

In 1923 the Dutch geophysicist Felix Meinesz had suggested, on the basis of his oceanic gravity surveys in the Pacific, that convection currents in the mantle could be dragging the crust into the mantle along the trenches, creating lower gravity values along those sites. This idea explained both the oceanic trenches and the low gravity. In 1928 Meinesz, along with Harry Hess of Princeton University, Maurice Ewing of Lehigh University, and Edward Bullard of Cambridge University, was invited to join gravity mapping expeditions in the Caribbean and the Gulf of Mexico. Their work confirmed Meinesz's theory that negative gravity anomalies (values lower than expected) were associated with ocean trenches. Their interpretation was that less dense oceanic crust was plunging down into denser mantle material.

In 1939, just as World War II was about to begin, Harry Hess formulated a model of crustal plates moving under the force of currents in the mantle. Hess drew on work by David Griggs at the University of California who had constructed a physical model with paraffin floating on a vat of oil. Drums rotating in the oil created currents and caused the paraffin on the surface to move in a manner suggestive of Earth's crust. Hess calculated that currents in the earth's mantle moved at a rate between 1 and 10 centimeters (0.4 to 4 inches) per year.

During World War II Harry Hess became captain of a transport ship carrying troops to landings in the Pacific islands. Part of his duty was to conduct continuous depth soundings of the Pacific ocean bottom using a newly developed instrument called a fathometer. This was used to create maps of depths and bottom topography. These maps had obvious military uses, but Hess found them useful for his scientific interests as well. He discovered sets of flat-topped mountains (seamounts) under thousands of feet of water, which he named guyots, after Arnold Guyot the first professor of geology and geography at Princeton. Hess interpreted the seamounts as wave-eroded extinct volcanoes that had formed in shallower water along the oceanic ridges. They appeared to him to have moved into deeper water, riding on the oceanic crust away from an active lava conduit. This conclusion grew from his earlier work in the Caribbean gravity surveys and has become an accepted explanation for guyots.

After the war, Hess continued his oceanographic research, and in 1962 he published an article proposing that in the early stage of Earth's formation, a single convection cell existed in the liquid material of its interior. As the earth continued to cool and differentiate into layers, convection broke into multiple cells as the core formed. He suggested that oceanic ridges formed over the rising columns of the convection cells. As the rising mantle material extruded onto the ocean floor, it cooled and formed a new segment of oceanic crust that spread laterally on either side of the ridge. Where the spreading ocean floors meet continental masses they would plunge downward and become part of the mantle rock again. Any land masses, like islands, riding along with the oceanic plate would be welded onto a continent as the oceanic plate dived back into the mantle. Geologists now feel that much of western North America consists of these masses that have been tacked on to the continent as the oceanic Pacific plate plunged (subduction) under the North American continent. Figure 12.1 illustrates the general scheme.

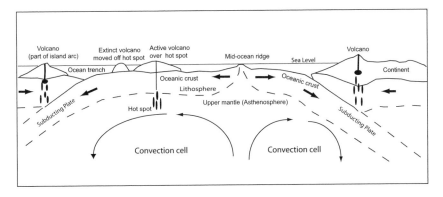

FIGURE 12.1

The lithospheric plates separate along a spreading center that
forms a midocean ridge. The convection cells in the mantle force
the plates to move laterally until they encounter another plate,
such as the continental plate on the far right. At that point the
oceanic plate subducts under the continental plate. Note
volcanoes forming over the zones of mountain building on the
leading edge of the continental plate. Volcanoes also form over the
hot spot, then move away as the plate spreads. (Adapted from
Kious and Tilling, 1996.)

Robert Dietz of the U.S. Coast and Geodetic Survey introduced the name
"seafloor spreading" for Hess's new model in which the position of a midocean ridge stays the same while the ocean widens outward in both directions. At this stage the majority of geologists were not ready to accept the idea
of a young ocean floor that was moving around. However, the possibility of
convection currents in the mantle was beginning to be accepted as a mechanism for moving the earth's crust. Interpretation of seismic data suggested that
the velocity of seismic energy through the mantle was characteristic of a solid
material. The idea later emerged that a material may behave as a solid under
short-term stress but over the long term it may have characteristics of a very
viscous fluid. An example of such a material is ordinary window glass, which
is rigidly solid but tends to distort and become wavy with flow lines under
the long-term steady pull of gravity.

In 1957 a topographic map of the seafloor, based on soundings, was made
by Marie Tharp and Bruce Heezen of Columbia University. This map provided more detail about the magnitude of midoceanic ridges than had ever

been seen. The striking features on these maps were the great extent of the ridges, which had been mapped only partially up to then. These ridges, sometimes called rises, may average 1,830 meters (6,000 feet) above the general level of the seafloor, and occasionally show on the surface as islands. The map revealed an almost continuous pattern of ridges over the earth, like stitching on a baseball, running through every ocean. They bisected the Atlantic and Indian Oceans and came together in the area between Africa and Antarctica, forming a global submarine mountain system 65,000 kilometers (40,000 miles) long. The summits of the ridges were marked by linear valleys along the crests of the oceanic ridges that are now regarded as the actual line of rifting between crustal plates. Molten material from the mantle appeared to have extruded from the linear valleys and formed new ocean floor as the crustal plates on either side moved away from the ridge.

Earlier geologists and geophysicists had rejected Wegener's hypothesis for lack of a plausible force to move continents. This new evidence for crustal movements drove scientists to accept currents in the mantle as the only plausible force acting on the crust. No other force was adequate. Today, currents in the mantle, though still unproven, are widely accepted.

Hess estimated that the entire ocean floor was replaced by new mantle material coming out of the ridges every 300 to 400 million years. He observed that this continual renewal of ocean floor would account for the relatively thin veneer of sediments on the floor and the lack of rocks older than Cretaceous (144 million years) in the oceans. On land, rocks older than Cretaceous are common, and the oldest known are nearly 4 billion years old. Beginning in 1968 the oceanographic drilling ship *Glomar Challenger* sailed in many parts of the world retrieving rock and sediment from the ocean bottom. The oldest rocks it found were 150 million years (Jurassic), only slightly older than those Hess knew about.

Maps of locations of seismic events showed that many of the world's earthquakes were focused along the midocean ridges, indicating that these were active lines of movements within the earth. The focus of earthquakes also matched the positions of known trenches and transform faults (such as the San Andreas fault) where plates are sliding laterally past one another. More than one hundred worldwide seismic monitoring stations, an outcome of the need to monitor adherence to the nuclear test ban treaty of 1963, provided seismic event maps that effectively showed the boundaries of each of the crustal plates of the world (figure 12.2).

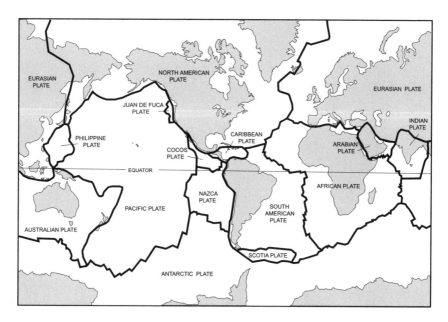

FIGURE 12.2

The global patchwork of plates consists of six large plates and
several smaller ones. The boundaries between plates are the focus
of earthquakes, volcanoes, ocean trenches, and mountain ranges.
(Adapted from Kious and Tilling, 1996.)

Interesting insights also began to emerge about the nature of Earth's magnetic field. Earlier studies by the French physicist Pierre Curie had shown that
heated rocks assume the magnetic polarity of the earth. He found that as a
rock cooled it retained the polarity acquired while it was hot, provided the
rock had first reached a critical temperature called the Curie point. It followed from this that igneous rocks, which had cooled from a molten state,
would have a magnetic polarity matching that of the earth, with north polarity of the rock oriented toward Earth's north magnetic pole. If the magnetic polarity of the earth changed, then the polarity of any rocks that were
molten at the time of the change would reveal the earth's polarity at that time.

Continental rocks collected from around the world and analyzed for magnetic polarity showed that Earth's magnetic field had changed position repeatedly through time. Rocks may have north polarity aligned with Earth's
present-day magnetic field, or they may have polarity oriented in almost any
other direction. The reasons for this lack of consistency with the present po-

larity of the earth could be either that the earth's magnetic poles had moved or that the land mass containing the rock had moved.

For many years the accepted answer was that the magnetic poles had wandered. This was the only plausible answer if the continents were considered static. As more evidence emerged supporting continental movements, the idea of polar wandering lost favor. That evidence came from collecting rocks of the same age (time constant) from different continents. If the continents had not moved, rocks of the same age would all orient to the same magnetic field. If the continents had moved, the rocks would not be oriented to the same magnetic field. Data collected in North America, Europe, India, and Australia showed the supposed paths of polar wandering to be different for each continent. This discovery in 1956 meant that the continents had moved and offered substantial support for the theory of continental movements.

As early as 1929 the Japanese geophysicist Motonari Matuyama had observed that the polarity of all lava flows younger than ten thousand years matched the earth's present magnetic field. Furthermore, all lava flows of Pleistocene age had reversed polarity. He concluded there had been a reversal of Earth's magnetic field about ten thousand years ago at the end of the Pleistocene. This work apparently went unnoticed by researchers in Europe and North America.[2]

Further studies of magnetic polarity of rocks showed that Earth's magnetic field had indeed reversed, with the positive pole oriented southward rather than northward. Evidence showed that Earth's polarity had reversed hundreds of times, including nine times in the past 4 million years. The average period between reversals was estimated to be about five hundred thousand years, and occasionally only a few thousand years. Radiometric rock-dating techniques provided the ages of rocks, resulting in a timetable of magnetic reversals. All this study of rocks on land set the stage for a major discovery derived from magnetic surveys in the oceans.

In 1963 an article in *Nature* appeared with the title "Magnetic Anomalies over Oceanic Ridges" written by Fred J. Vine and Drummond H. Mathews. In it they reported on the results of magnetic surveys by scientists in the North Atlantic, Antarctic, and Indian Oceans. In all these areas a similar pattern of magnetic stripes had been mapped. The pattern consisted of alternating bands of normal (north-oriented, as today) polarity and reversed polarity on the ocean floor. Furthermore, the pattern of magnetic bands appeared to form almost mirror images parallel to, and on either side of, an oceanic ridge. Because the patterns looked very distinct in some areas, but not so clear in others, some scientists first doubted the data. Perhaps some rocks had recorded the polarity accurately and others had not. Perhaps the polarity was dependent on what minerals were in the rocks. Many scientists, however, realized that a true breakthrough was at hand.

Vine and Matthews thought of trying to match their magnetic stripes on the seafloor with the known timescale of reversals of magnetic polarity. In this way they were able to determine the age of the seafloor along each of the magnetic stripes. They found that the seafloor was indeed as young as Hess had thought, and that the youngest parts were adjacent to the oceanic ridges. This became one of the pivotal papers in the development of plate tectonics theory. The magnetic stripes provided a means to date the seafloor in a manner similar to counting tree rings. Figure 12.3 illustrates this concept.

The development of magnetometers with the appropriate sensitivity and resolution was a stroke of good fortune. A magnetometer towed behind a ship at the surface was often 4,000 meters (13,000 feet) above the ocean floor. At such a great distance several conditions (polarity reversal rate, seafloor spreading rate, and depth of ocean) had to be optimal for the magnetic stripes on the

FIGURE 12.3
The periodic reversal of the earth's magnetic polarity forms stripes of alternating polarity on the ocean floor. The shaded stripes represent a period of polarity the same as at present. The unshaded stripes show reversals. As the plates separate, the cooled rock retains the polarity that existed at the time the magma extruded onto the ocean floor. The oldest event is labeled 1; the most recent, 6. (Adapted from Kious and Tilling, 1996.)

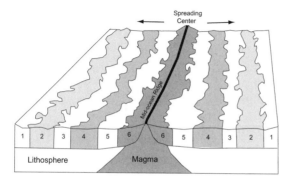

bottom to be detectable at the resolution of the magnetometers. The magnetic stripes had to be wide enough to be detected at the given resolution of the magnetometer. The width of the stripes is controlled by the rate of spreading and the frequency of magnetic reversals. These factors combined to produce stripes wide enough for the instruments to detect. If the spreading were slower and the reversals more frequent, the stripes would have been too narrow to be detected from the surface.

As more and more seafloor age data began to accumulate, geologists were able to calculate rates of seafloor spreading and develop a chronology of the ocean floor. Eventually the rates of spreading on each side of the ocean ridges in different parts of the world could be estimated. Depending on location, the rates range from 1.0 centimeters (0.4 inches) per year to 4.4 centimeters (1.7 inches) per year. Obviously the movement is only a tiny fraction of the rates (9 meters [29.5 feet] per year) that Wegener calculated in Greenland.

The time was right for another conference to discuss mobilism in light of the new information. Until then the earth science community was weary of the continental drift debate. The drift idea had been beaten to death, and there were no new discoveries or even new research ideas to discuss. Two meetings in the mid-1960s reopened the discussion. The question changed from "Have continents shifted?" to "How much and when?" Again the sites of the meeting were London and New York. The Royal Society of London convened in March 1964.[3] The program was simple. Seafloor spreading was not a major topic, and magnetic anomalies paralleling the ridge flanks were not discussed. Yet the tone of the meeting changed a number of fixists into mobilists. No doubt the attendees knew of the new developments and that influenced their openness to the presentations.

The most influential presentation was a map made by Edward Bullard, a highly respected geophysicist from Cambridge University, along with his colleagues J. E. Everett and A. G. Smith. The map was simply another fitting of the continents similar to Wegener's, except that it concerned only the continents bordering the Atlantic Ocean. Also, the match had been made by a least-squares-fit computer program. The program used the concept of Euler's theorem concerning the movements of objects on a sphere. They wrote the computer program using a statistical approach to match the irregular shapes of the continents of South America and Africa in one operation, and Europe and North America in another.

The best fit occurred at the 1,000-meter (3,281-foot) depth on the continental shelves. This depth could be considered to be the actual edge of the continents. Naturally, the shape of the continents at that depth is different from the shape controlled by today's sea level. So the continents on the map appeared somewhat distorted compared to more familiar maps. Actually the fit was close to that of Wegener's map, which some scientists had criticized as being greatly distorted. Minor gaps and overlaps along the continental edges were on the order of 30–90 kilometers (18–56 miles). The conclusion was that the fit was good enough to establish that such a fit would not occur by chance.

Why did this map from the Cambridge colleagues impress the audience and convert some skeptics? Wegener and several others had made similar attempts to fit the continents without impressing many scientists. The convincing element was the objectivity of a fit done by computer compared with that done by a biased person. One of the criticisms of Wegener's map had been that his continental distortions looked too much like a forced fit. Bullard and his associates were actually just refining the same effort and approach that Wegener had used. The other convincing factors were the general awareness of seafloor research under way and the acceptance of the possibility of convective currents in the mantle. Thinking that there was such a force made movement of continents seem more plausible.

Another line of evidence that helped convince the attendees at the 1964 Royal Society meeting was a project involving radiometric dating (done by measuring quantity of elements resulting from radioactive decay) of rocks in North America and Europe. J. A. Miller of Cambridge University divided rocks of those areas into six age groups and found that the groups matched across the North Atlantic in correlation to Bullard's map. Again the result was similar to Wegener's, but the method was more convincing.

Three years after the London meeting, another researcher, P. M. Hurley, used the same dating methods as Miller across the South Atlantic and matched rocks by age groups between Africa and Brazil. Hurley used Bullard's map to predict where rocks of a particular age would be found in Brazil that would match those in west Africa. His predictions were correct.[4]

The presentation made at the 1964 conference by Tuzo Wilson from the University of Toronto produced the main evidence related to seafloor spreading. He used fossils and radiometric dating to show that the islands of the

Atlantic were younger in the vicinity of the mid–Atlantic Ridge and older near the continental coasts. This suggested the islands were created at the ridge and moved outward.

The second conference of significance was held in New York City in November 1966, forty years to the month after the meeting of 1926 that had been so dismissive of Wegener's hypothesis.[5] The sponsors were the Goddard Institute for Space Studies and the Geology Department of Columbia University. The attendees included researchers on seafloor spreading and supporters of mobilism, as well as doubters who had some legitimate questions about the data presented. Much of the research presented was so recent that it had not yet been published, so doubters were seeing part of the evidence for the first time. Some of the papers dealt with completely new lines of evidence. There were twenty-five in attendance with seventeen presentations.

In the twenty years since the end of World War II there had been significant improvement of geophysical techniques, including better seismology, determination of heat flow through the crust, and more accurate echo sounding. In 1926 almost nothing was known of the ocean basins. Since then a window to the ocean floor had been opened. Also, recent research on land included a combination of new fieldwork and greatly improved rock dating that provided ages of the oldest rocks. The quantity and high quality of the data marked a new era in earth science research.

Summaries from the convinced and from the unconvinced were planned at the end of the conference. Edward Bullard gave his presentation summarizing the case for movement of ocean plates. The preponderance of evidence presented strongly favored crustal movement. The theme was that continents have moved great distances in the last 200 million years. Bullard admitted to skepticism in the beginning, but said he had been convinced by new studies in Australia on the shifts in the magnetic polarity in rocks.

Bullard pointed out that the reality of magnetic reversals was beyond doubt. The reversals have been worldwide and sudden.[6] Furthermore, he referred to convincing evidence by Vine regarding the global pattern of magnetic reversals laid out on the ocean floor parallel to the midocean ridges, with increasing age in both directions away from the ridges. The ocean floor is created at the axial rift valley on the ridge and magnetized with the earth's magnetic field at the time. He cited three independent lines of evidence supporting this: the field reversals of lavas on land, the same reversals found on the ocean floor, and the magnetic pattern of the ocean floor. The result is a timescale for the development of oceans and the rates of spreading of the oceans. The rate appears to be consistent throughout time with velocities from 1 to 6 centimeters (0.4 to 2.4 inches) per year depending on location.

Bullard summarized two ideas that could explain why the Atlantic Ocean opened. One said that the continents were carried on great crustal plates like a conveyor belt by convection currents rising under the midocean ridge and spreading to either side. The other said that the interior of the earth swelled and cracked the surface. However, such expansion would have required a 10 percent increase of Earth's radius in just the last 2 percent of geologic time, and it would lengthen the time of a day's rotation. There is no supporting evidence for either expansion or lengthening of the day.

Bullard commented on new information about convection currents in the mantle. "Great efforts have been made by engineers to obtain materials that do not creep at temperatures between 500° and 1,000° Celsius, but even the best of them creep at rates which would be significant to us."[7] The upper mantle near the crust would have to behave like a liquid, with a viscosity low enough to allow convection.

His thoughts for future work brought out questions that geologists have since confronted. Was the breakup of continents a one-time event in the Mesozoic or a stage in an ongoing process during which continents collided and later split apart several times? Are some older mountains, such as the Appalachians, the product of one of these earlier events? Were Asia and Europe forced together to form the Ural Mountains? Have continents been enlarged by accretion of land masses encountered in the path of their movement? Most of these insightful questions have since been decided in the affirmative by the work of many geologists.

At the end of Bullard's summary there were no dissenting voices. Gordon MacDonald had been scheduled to give a summary for the opposition but was unable to attend the final session. No one volunteered to speak for him. So Bullard rose to say what he thought MacDonald would have said had he been there. Not surprisingly, Bullard's view of the negative side was not convincing. His talk on MacDonald's behalf did not appear in the final proceedings.

Robert Phinney, a geologist at Princeton University, compiled the proceedings and wrote a useful introduction that expresses the tone of the meeting. The following comments are based on his introduction.

The idea of continental drift did not gain recognition as a serious hypothesis until the field of paleomagnetism brought Wegener's proposals back to life. Also, acceptance of convection currents in the mantle led to a new version of the debate on the earth's crust. New information about the phenomenon of creep in crystalline solids provided a legitimacy to the debate on convection. It gave a new view of the continental movement question and the debate between mobilists and fixists. Phinney identified Bullard's computer-fitted map in the London 1964 conference "as the most direct evidence for continental drift in the geological record."[8]

Phinney commented that earlier proponents of continental drift had over-interpreted some scanty information based on unproven assumptions. "It should be pointed out that much of the polarization of opinion over these issues was aroused by some of the liberties taken by enthusiasts of drift at the expense of established laboratory and field data."[9] The "liberties taken" must have been unproven assumptions and overinterpretation of weak data.

He found the most significant development to be the discovery of the reversals of the earth's magnetic field preserved in rocks, providing a timescale for crustal movements. Phinney referred especially to the discovery that lavas of a given age had consistently indicated the same positions for Earth's magnetic poles, regardless of location in the world.

After forty years, the concept of continental mobilism was at last accepted, though a decreasing number of dissenters persisted for some time. Great amounts of data from the seafloor had provided evidence far beyond Wegener's matching coastlines, fossils, geology, and paleoclimates. Wegener deserves credit for assembling multiple lines of evidence beyond any done before, but crucial information was not yet known. His approach was new and innovative. And though some of his interpretations were based on inadequate maps, Wegener's reconstruction of the continents around the Atlantic Ocean turned out to be close to Bullard's map of 1964.

Wegener's notion of Earth's rotation effects was so obviously inadequate that critics pounced on that point without considering the overall concept. His credibility problem was greater because of his lack of geologic background and experience, but his background probably accounted for his interest in the multidisciplinary approach. Wegener's hypothesis also presented problems with the physics of continents plowing through dense oceanic crust like a boat trying to move through mud. But had he not attempted to explain the reasons for continents moving, no doubt he would have been criticized for *not* identifying a force that could move continents.

Many phenomena are accepted without an understanding of their exact cause. The navigational compass was used long before anyone knew why it worked. The same could be said for gravity, electricity, light, and sound. The reason is that anyone can see the compass point north, or hear the sound of thunder, and have no doubt that the phenomenon exists. Wegener had a great disadvantage by being unable to show convincingly that continents had actually moved. Techniques available to him for geodetic measurements were too crude to measure the real distance of movement. The result was that he produced such large estimates of movement that North America could have moved completely around the world in the proposed available time. Yet he made it a part of his evidence because the estimated movement was in the desired direction. The margin of error in his calculations was greater than the actual movement.

Wegener's critics correctly identified the weakness of Wegener's idea, but they failed to discuss the core question of mobility vs. fixism. They missed the opportunity to consider the assumptions on which both ideas were based and look objectively at the weaknesses and strengths of each. Certainly the land bridge hypothesis that many of them held had serious weaknesses as well. A participant at the 1926 conference is quoted as saying, "If we are to believe Wegener's hypothesis we must forget everything which has been learned in the last seventy years and start all over again."[10] That statement exposes the rigid conservative mentality of an earth science establishment faced with a revolution in their discipline. While caution about accepting new ideas serves a useful purpose, perhaps some of the fixists should have considered starting over again.

By the time of the 1966 conference Wegener's critics had retired, and new scientists were on the scene. They were not mobilists at the beginning of their careers, but most were open-minded and willing to accept new ideas based on reasonable interpretation of data. And they had much more data than their predecessors. Their interpretation of the data led to a new mobilist theory that matched Wegener's in a general way, but was radically different in the details. The Earth was not expanding to create new oceans; there were no hidden masses of continental land bridges now sunken into the ocean floor; and mountains were not the result of global shrinkage. Instead of envisioning continents being shoved through the oceanic crust, they imagined large crustal plates consisting of ocean floor moving over the globe. Their data suggested plates being created along the ocean ridges and then being destroyed by subduction along the ocean trenches. The crust was being destroyed at the same rate it was being created. Now they accepted the probability of convection in the mantle as a driving force but were not concerned about proving it before stating their case for mobility.

The conferences of 1926, 1964, and 1966 marked major trials of Wegener's passion about continental drift. In the first he was vilified, but his confidence was not shaken. He felt that within ten years everyone would be convinced of the validity of his drift theory. In 1966 Wegener would have been eighty-six, and he no doubt would have felt pleased with the belated support of his ideas. Certainly, he would have been fascinated watching the scientific results unfold in the professional journals.

From the 1960s onward the earth science community rapidly began to accept the idea that continents had indeed moved laterally. The acceptance of lateral movement was helped when many unexplained phenomena could now be understood more clearly by applying the theory of mobile continents. When new theories explain phenomena that old theories cannot adequately answer, a scientific revolution is under way.

Today this revolution has led to widespread acceptance that Earth's crust is mobile and that continents have been joined and separated more than once. Most geologists believe the continents are not moving in isolation, but that each is a part of one of several great crustal plates that may include both ocean basin and land mass on a single plate. Furthermore, the movements and collisions between great crustal plates are the source of mountain-building forces that produced the earth's mountain belts. This change in thinking about the earth is revolutionary. It provides a unifying theory in earth science that answers many questions about Earth's history. There is such confidence in this theory that unanswered questions are assumed to be items that just need a bit more time and data before they can be understood.

# 13 ❖

## Progress after 1966

*Discovery consists of seeing what everybody has seen*

*and thinking what nobody has thought.*

—ALBERT VON SZENT-GYÖRGYI

### Greenland Expeditions after Wegener

Strong interest in Greenland continued through the 1930s. It was a new frontier that captured the imagination of nations with plans for future air travel, as well as adventurers wanting a new challenge.

In the same year that the Wegener expedition ended, the German pilot Wolfgang von Gronau became the first to fly across the Greenland ice cap—a distance of 1,368 kilometers (850 miles). Also, in 1931 Arne Høygaard and Martin Mehren, two Norwegian medical students, crossed the ice sheet just for sport. They went up the Kamarujuk Glacier to the West Station where Wegener's party was then located. The Germans and Inuits helped them haul two sleds and twelve hundred pounds of equipment up the glacier front. Their sled runners were covered with silver for easier sliding and to prevent sticking. They left on July 10 and reached the east coast on August 15, where they met a Norwegian ship for the return to Norway.

In the winter of 1932–1933 the University of Michigan along with Pan American Airways sent an expedition to gather data of interest to aviation over the area. In 1932 Ernst Sorge, who had spent a winter in Eismitte, returned as part of another expedition to continue glacial research begun during the Wegener expedition. During this time Sorge participated in flights that filmed the west coast of Greenland.

As a publicity event in 1933, Charles Lindbergh and his wife, Anne, flew across the ice sheet in an open cockpit seaplane, the *Tingmissartoq*. Their purpose was to study and promote the possibility of regular air travel from North America to Europe. They flew from New York to Labrador to Godthåb, on the west coast Greenland, and then made a short flight to Holsteinborg, Greenland, before flying 725 kilometers (440 miles) across the ice cap to Angmagssalik, Greenland. They continued on to Iceland, Europe, Africa, and South America. Five months later they returned to their starting point.

A French expedition led by Paul-Émile Victor crossed the ice sheet in 1936 with four men and three sleds traveling from west to east. As with many previous expeditions, this one became a race for survival. They threw away all excess freight in a mad dash for the east coast, collecting almost no scientific data.

One of the last expeditions of that decade was conducted by Oxford University in 1938. This group, led by P. G. Mott and J. C. Sugden, conducted glacial research on the Sukkertoppen Iskappe, an independent mountain glacier outside the main ice sheet on the west coast. Oxford had maintained a research station in this area for several years. By the end of the 1930s, research on the ice cap had come to a halt. The need for information about air routes over Greenland waned as airplanes became bigger and more powerful, with longer flying ranges. Also by 1939 world economic conditions were weak and the unstable political situation erupted into World War II.

## Greenland after 1940

Before 1940 most expeditions to interior Greenland, or any polar area, became a struggle for survival. Whether using horses or dogs it was only possible to transport slightly more than was consumed along the way. Men had the skills to survive the cold, but transportation capabilities limited the amount of food and fuel they could carry.

Just before World War II Richard Byrd demonstrated the advantages of using improved technology to solve some of the problems in polar research. He used airplanes for reconnaissance over large areas, looking for surface routes. In the process he flew over both poles. His Antarctic station, Little America, was better equipped than any station before that time. He gave particular emphasis to communication and had daily radio contact with the outside. In this way he could report any difficulties immediately and could also send information and data that allowed concurrent updating of maps back home. He was the first to use a tracked motor vehicle in the Arctic. These items all became the norm for later expeditions.

After World War II everything about mounting a polar expedition changed (figure 13.1). Cargo-carrying airplanes, large sleds with cargo and living quarters towed by heavy tracked vehicles, radios, packaged foods, improved fabrics for clothing: all these expensive innovations became available for one reason—the cold war. Suddenly the Arctic was an important frontier, with the Soviet Union and Western governments (Denmark, France, the United Kingdom, Canada, and the United States) all focused on surveillance and research of the Greenland ice cap.

Paul-Émile Victor left France when the Germans invaded and joined the American army. His Arctic experience landed him in Alaska, where he became acquainted with the capabilities of the weasel, an amphibious tracked vehicle developed by the American army for snow travel. After the war he organized and coordinated French expeditions in both the Arctic and Antarctic, introducing new procedures to improve the efficiency of polar expeditions. He began large-scale operations that could be applied to both north and south polar regions, sometimes even sharing equipment. If one project no longer needed a piece of equipment, it could be sent to another region. With the use of weasels, which were available cheaply as army surplus, and air transport for air drops, he was able to keep remote stations well provisioned. Victor also initiated in polar regions the use of radiosondes to obtain temperature and air pressure with instruments and a radio carried by a weather balloon.

Constant weather reports from remote stations became important for air travel to the stations and for setting times to make additional air drops of provisions. Thus modern expeditions became more expensive and could only be financed by national governments. Victor's first expedition in 1949 established Station Centrale, which was located on the same site as Eismitte. There they could begin from a known data point for comparison. At this time no expedition wanted for anything. Fuel was abundant enough to use for both light and heat at the stations, and provisions could be air-dropped in just one day of good weather.

FIGURE 13.1

Today's polar expeditions form trainlike convoys of supplies and portable living quarters with (*top*) hot showers; the convoys are pulled by (*bottom*) wide-track vehicles. (Photo (*top*) by Fristrup and photo (*bottom*) U.S Army, both from Fristrup, 1966.)

Despite the improved equipment, men in the Arctic still faced danger. On a return trip from Station Centrale, a weasel driver, Jens Jarl of Denmark (representing the Danish Ministry for Greenland on the expedition) pulled out of the convoy line, fell through an ice bridge over a crevasse, and disappeared, along with his passenger. Others came and saw the weasel upside down about 25 meters (80 feet) below, but the two men were not visible nor was there any answer to shouts. All the rescue gear was on that weasel so they had to radio for help and wait seven hours for an Icelandic plane to drop ropes and equipment. They found the two men dead deep in the crevasse, but could not retrieve the bodies, because the weasel overhead was on the verge of slipping further into the crevasse and endangering the rescuers.

More than twenty temporary ice stations were established in Greenland, beginning with Eismitte and the British station in 1930. Two of them (French, 1949, and American, 1954) occupied the same position as Eismitte, providing evidence of the value of Wegener's effort in Greenland.


## Glaciation and Glaciers

Since those early efforts at Eismitte to study glaciers in polar regions, many new discoveries have been made about the origins of continental glaciation. Much of this information is derived from new instrumentation for age dating and for chemical analysis. One such innovation involves the measurement of oxygen isotopes in sediments and shells.

An analysis of oxygen found in limestone and calcareous shells shows a variation with the salinity of the water in which the calcium carbonate formed. Oxygen with eight neutrons accounts for 99.76 percent of total oxygen in the environment. However, an increase in the number of neutrons to ten creates an isotope of oxygen that is rare, only 0.2 percent of total oxygen, but very important to the study of past environments. The concentration of the rare isotope is greater in seawater than in freshwater. Further, as seawater becomes less saline by the addition of freshwater, as from major melting of glaciers, the relative concentration of the heavier isotope decreases. The process of calcium carbonate formation in shells of sea animals extracts oxygen from the water, including both types of oxygen as available. Chemically, both types of oxygen behave the same, so if both are present each will be present in the calcium carbonate maintaining the same proportions as the seawater in which they formed.

The presence of additional freshwater in polar seas is an indicator of melting or increased rainfall associated with warmer temperatures. In this way oxygen isotope ratios provide a thermometer for past temperature variations. The climatic implication of this is that oxygen isotope ratios in fossil shells found in ocean sediment cores match the same cycles as Milankovitch's radiation curves—too well to be a coincidence. This phenomenon reawakened researchers awareness of a major role for solar radiation variation in glaciation.

The consensus now is that ice has spread in Europe and North America at times of cool summers and long, mild winters—as Köppen predicted. Mild winters can produce more snowfall because even slightly warmer air is able to carry greater amounts of moisture. During extended periods of long warm summers, glaciers have melted. Small changes in radiation received at a high latitude over a long time (100,000 years) make big changes in ice sheets. At last scientists had clear evidence that motions of the earth's orbit triggered the series of ice ages.

If orbital cycles of a few tens of thousands of years are controlling glaciation, as Milankovitch suggested, why was there glaciation in the Late Paleozoic, then a 300-million-year gap with no continental scale glaciation? The probable reason for this is that another condition must also be met: large land areas must be located in the high latitudes for glaciation to occur. Therefore, the drift of continents would have played a major role in determining that enough land mass was located in the high latitudes to make glaciation possible when the radiation cycles reached a low. If most of the land mass were in middle and low latitudes, there would be no glaciation even when the orbital changes were right for cooling in the high latitudes.

In the late Paleozoic the bulk of land areas were located in a large mass called Pangaea, and a portion of the mass extended into the low latitudes of the southern hemisphere. This was when Paleozoic glaciation occurred in Brazil, Argentina, South Africa, India, Antarctica, and Australia. The breakup of Pangaea caused large land masses, North America and Eurasia, to move into the northern hemisphere's high latitudes. Then these land masses became sensitive to astronomical variations. The result was cooler summers and milder, snowier winters on large land masses in high latitudes. Then the glacial cycles Milankovitch devised (100,000 years, 41,000 years, and about 23,000 to 19,000 years) began showing up clearly over the past 600,000 years. Hence, glaciation appears to depend on astronomic cycles of radiation plus the right distribution of large land masses.

As the Greenland ice cap increases in thickness, it influences the land in several ways. The sheer weight of an ice mass 3,048 meters (10,000 feet) thick causes Greenland to sink significantly into the earth's mantle—a phenomenon called isostasy. The same happens when weight is added to a boat on water. The Greenland land surface under the ice has a basin shape that is pushed to near or slightly below sea level. In a counteraction the mountains at the east and west margins have risen slightly. Imagine sitting on an air mattress when another person sits down—they sink and you rise. The concept of isostasy shows also that the material under the earth's crust is not rigid, but has plasticity, which is cru-cial for lateral movement of crustal plates. Figure 13.2 shows Greenland through a sequence of changes beginning with no ice, to a period of maximum ice ex-tending beyond the land, and present-day conditions.

Another important feature of the Greenland ice cap is the flow of ice east-ward and westward from a north-south axial line running most of the length of the island. The movement is more complicated than lateral only; com-paction and downward movement occur as well. A layer of ice becomes com-pressed and thinner, creating a downward component to the outward flow of

FIGURE 13.2

(*a*) Before an ice cap formed, Greenland would have been well above sea level. (*b*) During maximum glaciation, ice extended beyond the land as an ice shelf, and the weight of ice depressed the land; sea level was lower. (*c*) At the present time, the ice cap is somewhat smaller and the land remains depressed. (Adapted from Fristrup, 1966.)

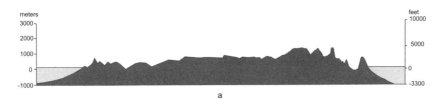

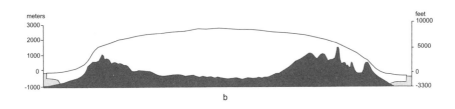

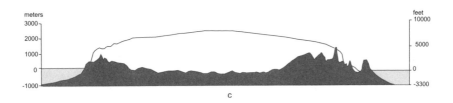

the ice. Eventually ice reaches the coastal mountains, and in many places the ice is extruded toward the ocean along valleys between the mountain ridges.

At this point the ice cap changes into a series of valley glaciers similar to those found in the Alps, Northern Rockies, and many other mountain ranges. As the ice scrapes and plucks rock, the valley is transformed into the characteristic U-shape of mountain glaciation. Where the U-shaped valley extends into the sea, a big inlet called a fjord is formed as sea level rises. In other places, the ice cap is not so obstructed by mountains along the coast and forms a continuous, insurmountable ice wall right to the edge of the land. Explorers attempting to reach the top of the ice cap chose to scale the relatively easier slope of the valley glaciers in the fjords.

Ridges of rock debris, called moraines, form at the sides and leading edge of glaciers. Wegener's party began moving supplies for the West Station and Eismitte up a trail cut into the ice of the Kamarujuk glacier front. When that trail melted away during the summer thaw, another trail along a lateral moraine at the edge of the glacier became the route up to the West Station. A terminal moraine marking the maximum advance of the Kamarujuk glacier is probably buried underwater near the mouth of the fjord.

As glacial ice starts down steeper slopes toward the sea, stress fractures create crevasses in the ice mass. All explorers to Greenland experienced the hazards of crossing crevasses at the edges of the ice cap. Once past these hazards, travelers found the rest of the ice cap essentially free of crevasses.

Current studies of the Greenland ice cap show a marked retreat of ice around the margins. This is due to an increase in the rate of melting, and the probable cause is global climate change, driven by natural processes and human activities. Photographs of the Kamarujuk glacier, which Wegener's party climbed, show the summer 1931 position of the glacier and the retreated position in autumn 2001 (figure 13.3). This effect is apparent over all of Greenland in satellite images.

## Summary of Plate Tectonics

With an awakening to the possibilities of plate tectonics, there began an ongoing surge of research and reinterpretation. A brief summary of current thinking about plate tectonics and mountain building is included

FIGURE 13.3

Photos of the Kamarujuk glacier show the place where Wegener's
expedition began its ascent: (*top*) in summer 1931 the glacier
extended to the bottom of the slope; (*bottom*) in September 2001
the glacier shows marked recession. This is partly due to the later
time in the season and partly due to a general retreat of the ice
over the years. (Photo (*top*) used with permission of the Alfred
Wegener Institute for Polar and Marine Research, Bremerhaven,
Germany; photo (*bottom*) courtesy of Clare Dudman.)

here. Many books covering the subject in greater detail are available for readers wanting to know more. Jon Erickson's book *Plate Tectonics: Unraveling the Mysteries of the Earth* provides an interesting general account of the current concepts and research results. *Continents and Supercontinents*, by Rogers and Santosh, contains a wealth of more advanced material.[1]

The concept of seafloor spreading developed quickly into a global view of plate tectonics. A rapid growth of interest and new research induced most earth scientists to interpret their work in light of the new concept. Alfred Wegener's continental drift was transformed into a new concept of mobilism that has revolutionized the way geologists and geophysicists think about every problem they encounter.

The concept of plate tectonics assumes that the earth is covered by large crustal plates matched together like giant flagstones. The plates are set in motion by the force of convection currents in the upper mantle. Continents are part of most large plates, but some plates are oceanic crust only. When continents are part of a plate, they move with the plate as a unit.

The edges of plates are sites of most of the earthquake and volcanic activity in the world. One type of plate edge is along the midocean ridges where mantle material extrudes, creating new seafloor. Another type of plate edge is found at the deep ocean trenches where a plate subducts back into the mantle. The third type is the collision of two plates moving either head-on (the India plate moving into south Asia creating the Himalayas), or grinding past each other laterally (the Pacific plate moving northwest along the North American plate forming the San Andreas Fault).

Midocean ridges are lines of tension and rifting where new material is added to the edges of plates as they spread away from the ridge. One researcher estimated the amount of basalt extruded from the rift to be about 50 billion tons per year. As the new crust material cools below the Curie temperature, the polarity of the earth's magnetic field is locked into the rock. Volcanoes may develop along the ridges and eventually reach the surface, appearing as islands. Iceland is one example of such an island that is on the ridge and still volcanically active. Some Atlantic islands appear to have remained attached to one of the plates and moved away from the ridge with the plate.

Other volcanic areas occur along the zone of subduction. As the oceanic crust plunges deep into the mantle at one of the deep ocean trenches, the crust melts. Being less dense than the surrounding mantle material, the melted crust tends to rise and emerge violently as volcanoes. The entire rim of the Pacific Ocean is being consumed by advancing continental plates, creating a belt of volcanoes and earthquakes sometimes called the Pacific Ring of Fire.

The first map of plates was published in 1968. On it six large plates and several smaller ones were defined by areas of minimal seismic activity, bounded by lines of high seismic activity. The global seismic network made it possible to create such a detailed map. More recent maps of global plates are only slightly modified from the first version (see figure 12.2).

Where volcanic activity is not located near plate boundaries, such as the Hawaiian Islands, the cause is a persistent hot spot in the mantle, which in this case sent lava through the Pacific plate and built a mountain of lava that reaches far above sea level. As the Pacific plate moved toward the northwest, the volcanic island moved with it and became inactive, but the hot spot in the mantle stayed in the same location. The hot spot then has to begin all over again making a new island. Each island in the entire Hawaiian chain has been over the hot spot at one time, and as it moved away volcanic activity stopped. The oldest island, Kauai, is the most northwesterly one and is about 5 million years old. The youngest is the big island Hawaii, which has taken less than 1 million years to reach its present size and is still active. Just offshore, at the southeast edge of the big island, is a new underwater volcano that rises 3,350 meters (11,000 feet) from the ocean floor, but is still 975 meters (3,200 feet) from the surface. Eventually, in about 10,000 years, it will reach the ocean surface and become a new island. Already it has been given the name Loihi.

Extending northwestward beyond Kauai is a chain of seamounts, each of which was over the hot spot at one time. This feature, called the Hawaiian ridge, forms an underwater continuation of the Hawaiian Islands, taking an elbow bend northwestward about 3,219 kilometers (2,000 miles) from Kauai where its name changes to Emperor Seamounts. At its northerly extent the material in the seamount is more than 60 million years old and reaches almost to the outer tip of the Aleutian Islands.

Yellowstone Park is believed to be over the site of another hot spot. As the North American plate has moved westward, lava from that hot spot extruded onto the surface leaving great basaltic lava beds in what is now eastern Oregon, Washington, and the Snake River Plain of southern Idaho. The Craters of the Moon National Monument in Idaho is an excellent place to view the results of that area's pass over the hot spot from fifteen thousand until as recently as two thousand years ago. In the next stages of movement, the North American plate may shift so that South Dakota is in a position over the same hot spot, with resulting lava flows and thermal springs. The remnants of hot spots, such as the Hawaiian Islands or western lava beds, are good indicators of the direction of plate movements.

Continents also have rifting zones that are in the process of splitting apart. Continental rifts formed the rift valleys of east Africa where great herds of animals roam the grasslands. An extension of the east African rift continues northward and forms the Red Sea, which appears to have started as a continental rift and is becoming wider by about 2 centimeters (0.8 inches) per year as rifting continues. However, continental rifts are different from oceanic rifts. They lack central ridges extruding lava and also lack magnetic stripe anomalies on the valley floors. Another zone of continental spreading is the vast area of western North America extending from Salt Lake City, Utah, to Reno, Nevada, and from eastern Washington to southern Arizona and northern Mexico. The basin and range character of this region is the product of continental rifting. Within the past 15 million years Reno and Salt Lake City have moved farther apart by 320 to 480 kilometers (200 to 300 miles) averaging a rate of about 2.5 centimeters (1 inch) per year.

Where plates converge, one of several results may occur. If both plates are oceanic with no continent, one plate will disappear under the other (subduction) forming an ocean trench as it sinks into the mantle. Volcanic island arcs arise on the overriding plate, such as Japan, which located on the Eurasian plate as it overrides the Pacific plate that is plunging back into the mantle. The East Indies and the Aleutian Islands are other examples of island arcs in the Pacific. If one plate has a continental mass (like the west edge of the South American plate) it will override the oceanic plate (Nazca plate) and a trench will form along the line of subduction. The continental margin will be compressed into a range of mountains with volcanoes such as the Andes.

When both colliding plates are continental it is likely that one will override the other—not subduction, but a thrusting of one plate over the other. An example of this is the so-called overthrust zone of Nevada, Utah, and western Wyoming, which were once near the western edge of the North American plate while the Pacific plate overrode. Another option when plates collide is a crumpling and folding of rocks on both plates. In either case the result is crustal shortening. Examples of this type of collision are the raising of the Alps as Italy comes shoving northward, or the Himalayas rising as India moves north. As two continents come together, an ocean disappears and the two land masses become welded, or sutured, together. This appears to have occurred when North America pushed into Europe to form the Appalachian Mountains during the assembly of Pangaea throughout the Paleozoic.

Many lines of evidence support the theory of crustal mobility today. It is widely accepted as fact by most geologists without direct evidence of the mechanism driving it. If convection currents in the mantle provide the moving force—the most accepted theory—clear proof of their existence is still not available. Hence, the driving mechanism problem that repeatedly plagued Wegener is still not solved, but convection is accepted for now as the only plausible explanation for movements of massive blocks of the earth's crust. In the meantime the widespread acceptance of crustal mobilism has opened the door to revolutionary discoveries about the history of the earth. The most striking aspect of plate tectonics as an underlying theory behind every geologic study is the awareness that many explanations for past events are now tied to the same concept. No longer are field studies isolated cases trying to solve a local problem. Rather, they are all tied together by a unifying theory of crustal movements. A few geologists dispute some of the details of plate tectonics and point out that their own fieldwork does not support the theory. Such discrepancies bear serious notice. Certain events and processes truly have only local effects and are not tied to global plate movements. Also, work that disputes the theory shows that certain details worked out by others regarding the geologic history of an area may need reinterpretation. Despite some disagreements, most local phenomena can be tied to plate tectonic theory. It has become the central concept for geology and geophysics today.

Geologists now say that plate movements have been a part of the earth's history since the beginning, long before the breakup of Pangaea. In the Precambrian, for example, the crust first began differentiating into land masses and ocean basins. Rocks of oceanic crust pushed up onto land are as old as 3.6 billion years. This occurrence is best explained by plate collisions thrusting one type of rock onto another. Mountains resulting from continental collisions appear to have formed at roughly 400- to 500-million-year intervals. Prior to the formation of Pangaea in the Paleozoic, other supercontinents seem to have existed at 650 million years and 1.1 billion, 1.7 billion, and 2.1 billion years. Eventually, the continents could again rift apart and later re-form into supercontinents in a repeating cycle. They take about 160 million years to reach their greatest dispersal, and another 160 million to regroup into a new supercontinent. The new mass may last another 80 million years before it too begins rifting and separating in the dance of the continents. Hence, Wegener's hypothesis conforms to the crucial principle of uniformitarianism. The process of today indeed occurred continuously in the past. Geologists have been able to suggest a likely chronology of earth-moving events (table 13.1).

TABLE 13.1 *Geologic time chart*

| Era | Period | Years ago (millions) | | Related events |
|---|---|---|---|---|
| Cenozoic | Quaternary | | Pangaea Dispersing | Ice age (Pleistocene) with four ice advances ended—11,000 years ago |
| | | 2 | | |
| | Tertiary | | | |
| | | 65 | | Rockies, Andes Mtns. formed —— |
| Mesozoic | Cretaceous | | Pangaea Dispersing | |
| | | 144 | | |
| | Jurassic | | | |
| | | 208 | | |
| | Triassic | | | |
| | | 240 | | Appalachian, Ural, and —— Scandanavian Mtns. |
| Paleozoic | Permian (Pennsylvanian) | 286 | Pangaea Assembling | Glaciation in Africa, So. America, India. Coal in N. America, Europe |
| | Carboniferous (Mississippian) | 360 | | *Mesosaurus* reptiles and *Glossopteris* ferns lived |
| | Devonian | | | |
| | | 408 | | |
| | Silurian | | | |
| | | 438 | | |
| | Ordovician | | | |
| | | 505 | | |
| | Cambrian | | | Rodinia rifted apart |
| | | 570 | | |
| Precambrian | Archeozoic | | | Landmass of Rodinia formed |
| | Protozoic | | | |
| | | 3,800 | | |

Source: A. Wilson, *Compilations of Various Geologic Time Scales.*

In the late Precambrian (about 1 billion years ago) the land masses collided, creating a new supercontinent called Rodinia. This collision raised a 4,800-kilometer-long (3,000-mile-long) range of mountains along the suture of the two land masses. The mountains have long since eroded away, but the folded remnants are present in the crusts of eastern North America and western Europe. Near the end of the Precambrian (570 million years ago) Rodinia began rifting apart forming an ocean called the Iapetus Sea. Near the present-day Appalachians a belt of extinct volcanic activity is similar to that found along a midocean ridge that marks the site of Rodinia's rifting apart. The Ordovician period (438 to 505 million years ago) saw the continents reaching their maximum separation.

In the Paleozoic era, from the Devonian into Carboniferous periods (400 to 300 million years ago) continental masses again converged into a super-continent—Wegener's Pangaea. The assembling of Pangaea caused collisions that created some of the older mountain ranges, such as the Urals where Asia and Europe sutured together. The Appalachians, the Pennines in Britain, and the mountains of Scandinavia also formed during Paleozoic continental collisions. The Appalachians, based on their structure, could have been more than 6,100 meters (20,000 feet) high and have had ample time to erode down to their present size. As the land masses of Pangaea came together, the Iapetus Sea closed forever. When they split open again the Atlantic formed.[2] These collisions were spread over most of the Paleozoic era, creating mountain-building events of several different ages.

Then in the early Mesozoic (245 million years ago) the separation began and is still in progress today (figure 13.4). The Atlantic Ocean and the Indian Ocean opened as new water bodies. The Pacific began its continuing process of becoming smaller as continents moved in from both east and west. North America and South America began overriding the Pacific plate, raising the Cordilleran mountain system extending from Tierra del Fuego to Alaska.

As North America moved westward it rammed into large islands that became attached to the continent. The west edge of the continent had at one time been a roughly north–south line through what is now central Nevada. Through successive collisions land was added until finally the west coast formed at its present location. Each of the collisions with land from offshore caused the mountain ranges of western North America to buckle up. At the end of the Mesozoic, the Rockies, and more recently the Sierra Nevada, were formed.

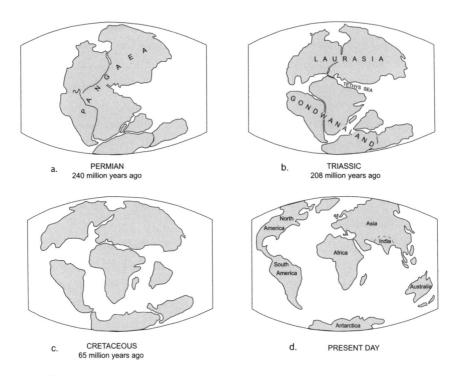

a.  **PERMIAN**
    240 million years ago

b.  **TRIASSIC**
    208 million years ago

c.  **CRETACEOUS**
    65 million years ago

d.  **PRESENT DAY**

FIGURE 13.4

The dispersal of Pangaea from a supercontinent at the end of the
Paleozoic to the present pattern of six major land masses covers a
period of 240 million years. (Adapted from Kious and Tilling,
1996.)

About 45 million years ago India pushed into Asia, raising the Himalayas. A
subducted portion of the plate raised the Tibetan plateau. Some 30 million years
ago Africa shoved into Europe, raising the Alps and Dolomites, and a continuation of the same motion later raised the Pyrenees and Atlas Mountains.

When major disasters occur, we get reminders that the plates continue to
collide. Mount Saint Helens in 1980, the 2004 tsunami in Indonesia, and the
Iranian earthquake in 2005 may grab our attention, but smaller events detectable only by seismograph occur daily. Today's technology makes it possible
to measure plate motion accurately. To measure the tiny annual movement requires making highly accurate determinations of location over distances of at
least one thousand miles. This can be done with the Satellite Laser Ranging
Program, which derives accurate measurements by laser impulses from ground

to satellite and back. Satellites also provide the Global Positioning System (GPS) capable of a precision of 2.5 centimeters (1 inch) between points 500 kilometers (310 miles) apart. Although this amount of error might mask continental movement of a single year, over a longer period of displacement it would be a minor error.

Another measurement method called Very Long Baseline Interferometry (VLBI) monitors radio signals from quasars (collapsed stars). The difference in signal arrival time at various stations on Earth determines the distance between the receiving stations. This method is the most accurate and can achieve a few parts per billion, which is the equivalent of an error the width of a human hair in a measurement of 3 kilometers (almost 2 miles). Measurements of this accuracy can easily detect continental movements. The results of these direct measurements agree with estimates based on geologic analysis of the spacing of magnetic stripes of the ocean floor.

So the direct evidence that eluded Wegener has become available more than sixty years after his death. Du Toit and a few other scientists embraced Wegener's proposal, but the geologic establishment as a whole waited more than fifty years after Wegener's first publication to accept the idea of horizontal continental movements.

After that much time, is it reasonable to consider Alfred Wegener the initiator of continental movement? The researchers of the 1950s were not testing his hypothesis. Their discoveries were independent of anything Wegener had done. Yet, as the significance of their results began to dawn on them, occasionally they commented that Wegener was right after all. Though plate tectonics is a much more sophisticated concept than continental drift, Wegener's idea of Pangaea, lateral movement rather than land bridges, and certain of his climatic, paleontologic, and geologic lines of evidence have survived. He certainly pointed to a new direction regarding the cause of mountain ranges. He provided an alternative to the inadequate idea that mountains were formed by contraction of a cooling Earth. When the discoveries that led to plate tectonics were made in the 1950s, researchers found that Wegener's hypothesis and extensive evidence had indeed been a precursor. For this some geologists have dubbed him the Father of Continental Drift.

In addition to his fame for the theory of continental drift, Wegener pioneered several ideas in meteorology, particularly his seminal suggestions on the theory of raindrop formation. The means by which ice crystals in clouds attract supercooled water droplets is now called the Wegener-Bergeron-Findeisen process. Unusual halos caused by ice crystals in the cold air above an ice cap also bear his name.

Alfred Wegener's ability as a capable, creative scientist led to his ideas on continental drift, as well as other groundbreaking scientific studies in the Arctic. He faced difficult tests in connection with both endeavors—his hypothesis and his scientific adventures. He was a man of considerable imagination and physical endurance who sought to test his limits in both.

Alfred Wegener's image has come a long way since its nadir of the 1920s. In the middle of the twentieth century he was still viewed as a crackpot scientist by geologists clinging blindly to a failed paradigm. Today Wegener's legacy enjoys greater esteem. He holds the respect of many for his creative insights and his dogged efforts in the face of a hostile or indifferent earth science community.

Some recognition came to Wegener during his life for his contributions, but only for his work as a polar explorer. He received medals of service from Denmark for his first two Greenland expeditions. German scientific organizations also awarded him medals for his work in Greenland. After his death the city of Graz, Austria, renamed a street Wegenergasse. Hamburg named streets for Wegener, Köppen, and Georgi. The University of Graz installed a bronze bust of Wegener in the main building on campus. The university also donated a plaque honoring Wegener, which they had placed at the site of the expedition's West Station. The Danes gave the name Alfred Wegener Peninsula to the ridge beside the fjord where the expedition began. Today a prominent research organization called the Alfred Wegener Institute for Polar and Marine Research operates in Bremerhaven, Germany. Museums in Zechlinerhütte and Neuruppin, Germany, contain memorabilia and photographs from the Wegener families.

Wegener has joined the company of other notable innovators who broke old paradigms of thinking but made errors in certain details. Their critics rejected an entire concept because of discrepancies in details. The idea first proposed by Copernicus needed Galileo, Kepler, and Newton to get the details right. It is my belief that Wegener should be regarded in the same category as thinkers like Copernicus who had the imagination to see a new direction for science and a new way to think about answers to many questions.

 *Notes*

CHAPTER I

1. The term "drift" was not the best translation of Wegener's term, *Verschiebung*. A better translation would be "displacement." Drift caught on through common usage.

2. Else Wegener, *Alfred Wegener: Tagebücher, Briefe, Erinnerungen* (Wiesbaden: F. A. Brockhaus, 1960), 65.

3. Ibid., 66.

4. Otto Krätz, *Goethe und die Naturwissenschaften* (Munich: G. D. W. Callwey, 1992), 188–94.

5. Ulrich Wutzke, *Der Forscher von der Friedrichsgracht* (Leipzig: F. A. Brockhaus, 1988), 87.

6. Wegener, *Alfred Wegener,* 139.

7. The American geologist/geomorphologist G. K. Gilbert had proposed in 1892 that lunar craters were formed by meteoric impacts. He presented the idea in his address to the Philosophical Society on the occasion of his retirement as its president. The publication, G. K. Gilbert, "The Moon's Face: A Study of the Origin of Its Features," *Bulletin of the Philosophical Society* 12 (1893): 262–64, was read by few astronomers, and it is unlikely that Wegener was aware of it.

CHAPTER 2

1. Ulrich Wutzke, *Der Forscher von der Friedrichsgracht* (Leipzig: F. A. Brockhaus, 1988), 92.
2. Ibid., 96.
3. The Mid-Atlantic Ridge had been discovered by soundings in the 1870s by the H.M.S. *Challenger* in a three-and-a-half-year research voyage of soundings with samples of water, sediment, plants, and animals. The global extent of the network of oceanic ridges was not known until after 1950 when results from many voyages were compiled.
4. Ursula B. Marvin, *Continental Drift: The Evolution of a Concept* (Washington, D.C.: Smithsonian Institution Press, 1973), 63.
5. W. A. J. M. van Waterschoot van der Gracht, ed., *The Theory of Continental Drift: A Symposium* (New York: American Association of Petroleum Geologists, 1928), 75.
6. In 1981 an article by S. M. Totten titled "Frank B. Taylor, Plate Tectonics, and Continental Drift," appeared in the *Journal of Geologic Education*. In it Totten examined the text of both Taylor and Wegener and concluded that Taylor's work had indeed influenced Wegener.
7. Wegener called the continental crust SIAL for its predominance of silica- and aluminum-bearing rocks. The denser oceanic crust was named SIMA for its predominance of rocks containing silica and magnesium.
8. Ten years after Wegener's death, Wladimir Köppen, age ninety-three, wrote some supplements to the paleoclimate book he and Wegener had first published in 1924. In mid-June 1940 he sent this telegram to his publisher, "Please send proofs at once, I am dying." The page proofs arrived soon, but his daughter, Else, had to edit them, as Köppen died on June 22. Else Wegener-Köppen, *Wladimir Köppen: Ein Gelehrtenleben für die Meteorologie* (Stuttgart: Wissenschaftliche Verlagsgesellschaft M. B. H., 1955), 159.

CHAPTER 3

1. Ursula B. Marvin, *Continental Drift: The Evolution of a Concept* (Washington, D.C.: Smithsonian Institution Press, 1973), 83. The British Association for the Advancement of Science also met in 1922 to discuss and debate continental drift.
2. Ibid.
3. Ibid.
4. Ibid., 85.
5. Ibid.
6. Ibid.
7. Ibid., 86.
8. W. A. J. M. van Waterschoot van der Gracht, ed., *The Theory of Continental Drift: A Symposium* (New York: American Association of Petroleum Geologists, 1928).

9. Ibid., 97.
10. Ibid., 75.
11. Ibid., 140.
12. Ibid., 145.
13. Ibid., 87.
14. Ibid., 220.
15. Arthur Holmes, "Continental Drift," *Nature* 122 (1928): 431–33.
16. Arthur Holmes, "Radioactivity and Earth Movements," *Transactions of the Geological Society of Glasgow* 18, part 3 (1931): 559–606.

CHAPTER 4

1. Johannes Georgi, *Mid-Ice* (New York: E. P. Dutton, 1935), 2.
2. When glaciers calve, large pieces of the front edge break away, fall into the water, and float away as icebergs. Small boats nearby can capsize from the sudden waves generated by calving.
3. Hagen Schulze, *Germany: A New History* (Cambridge, Mass.: Harvard University Press, 1998); Henry M. Pachter, *Modern Germany, A Social, Cultural, and Political History* (Boulder, Colo.: Westview Press, 1978).
4. To help keep scientific research alive in the 1920s, German universities and learned societies pooled resources to create the Emergency Aid Committee for German Science (Notgemeinschaft der Deutschen Wissenschaft) to provide research funds to worthy German scientists of the day.
5. The intermediate stage of compaction and recrystallization as snow transforms into glacial ice is called firn.
6. W. Köppen and A. L. Wegener, *Die Klimate der Geologischen Vorzeit* (Berlin: Bonntraeger, 1924).
7. J. Imbrie and K. P. Imbrie, *Ice Ages: Solving the Mystery* (Short Hills, N.J.: Enslow Publishers, 1979), 98.
8. Place-names are those used on 1930 maps. Recent maps show different spellings for the Inuit names and change Danish names to the Inuit names in local usage. Present-day usage is shown in parentheses.
9. A *nunatak* is a mountain peak protruding through a glacier.
10. Else Wegener, *Alfred Wegener: Tagebücher, Briefe, Erinnerungen* (Wiesbaden: F. A. Brockhaus, 1960), 206.

CHAPTER 5

1. Else Wegener, ed., *Greenland Journey* (London: Blackie and Son, 1939), 16.
2. The Greenlanders Wegener refers to are Inuits, though many have Danish names and ancestors dating from the earliest Danish colonization.

3. Else Wegener, ed., *Greenland Journey* (London: Blackie and Son, 1939), 19.

4. Pemmican is made from dried meat pounded into a powder and mixed with hot fat and dried berries or fruit, if available. It can be molded into a loaf or canned for transport. The Wegener expedition had crates of canned pemmican that they either sliced and fried or made into soup. It could also be eaten straight from the can. Comments from some of the men inexperienced in polar exploration suggest that pemmican is an acquired taste.

5. Else Wegener, ed., *Greenland Journey* (London: Blackie and Son, 1939), 26.

6. At 71 degrees north latitude the sun never sets between May 17 and July 28. It follows a nearly circular path in the sky, lower in the north and higher in the south. At the June solstice the noon sun is in the south at 42 degrees above the horizon. Compare this with a noon solar angle of 73 degrees at the latitude of Philadelphia (40 degrees N) on the same date. At midnight the solstice sun is shining from the north only 3 degrees above the horizon at the latitude of West Station.

7. Else Wegener, ed., *Greenland Journey* (London: Blackie and Son, 1939), 40.

8. Lateral moraines are ridges of crushed rock debris formed along the lateral edges of an alpine glacier.

9. The east coast of Greenland at 71 degrees latitude is usually locked in ice until August, and is ice free only until mid-October. Because of a warm ocean current, the west coast is usually open from May into January.

10. Else Wegener, ed., *Greenland Journey* (London: Blackie and Son, 1939), 47.

11. Ibid., 49.


CHAPTER 6

1. The British Arctic Air Route Expedition also spent that winter gathering meteorological data. One man spent the winter in the middle of the ice cap about 485 kilometers (300 miles) south of Eismitte. He also had no radio.

2. Johannes Georgi, *Mid-Ice* (New York: E. P. Dutton, 1935), 82.


CHAPTER 7

1. The reindeer is a subspecies of caribou. The term reindeer is usually associated with the animals of northern Europe, but, for general usage, the name is interchangeable with caribou in North America.

2. Else Wegener, ed., *Greenland Journey* (London: Blackie and Son, 1939), 172.

3. Gontran de Poncins, *Kabloona* (Alexandria, Va.: Time-Life Books, 1941), 59.

4. Johannes Georgi, *Mid-Ice* (New York: E. P. Dutton, 1935), 117.

5. Ibid., 118.

6. Ibid., 120.

7. Ibid., 121.

8. This account of Wegener's attempted return trip is based on known locations of the abandoned sled, some campsite locations, Wegener's gravesite, and Villumsen's first two campsites. Wegener's notes were missing so the daily distances and the number of days traveled are estimates. The actual fate of Villumsen is not known.

CHAPTER 8

1 Else Wegener, ed., *Greenland Journey* (London: Blackie and Son, 1939), 113.

2 Ibid., 117.

CHAPTER 9

1. Johannes Georgi, *Mid-Ice* (New York: E. P. Dutton, 1935), 104.

2. Ibid., 107.

3. Ibid., 125.

4. Gontran de Poncins, *Kabloona* (Alexandria, Va.: Time-Life Books, 1941), 298.

5. Ibid. 137.

6. Ibid., 138.

7. Barry Lopez, *Arctic Dreams* (Toronto: Bantam Books, 1986), 216.

8. Johannes Georgi, *Mid-Ice* (New York: E. P. Dutton, 1935), 114.

9. Ibid., 124.

CHAPTER 10

1. Else Wegener, ed., *Greenland Journey* (London: Blackie and Son, 1939), 168.

2. Ibid., 200.

3. Johannes Georgi, *Mid-Ice* (New York: E. P. Dutton, 1935), 184.

4. Ibid., 202.

5. Ibid., 214.

CHAPTER 11

1. Johannes Georgi, *Mid-Ice* (New York: E. P. Dutton, 1935), 245.

2. Ibid., 242.

3. Ibid., 245.

4. Else Wegener, *Alfred Wegener: Tagebücher, Briefe, Erinnerungen* (Wiesbaden: F. A. Brockhaus, 1960), 253.

5. Ibid.

6. The Wegeners' eldest daughter, Hilde, died in 1936, and their youngest, Char-

lotte, died in 1989. Their middle daughter, Käthe, lives in southern Germany near Stuttgart. Else, the Köppens, and Alfred's siblings are all buried in the village of Zechlinerhütte, near Berlin. A Wegener memorial and small museum are also located there.

CHAPTER 12

1. Tectonics is the study of large-scale structural features in the earth's crust.
2. M. Matuyama, "On the Direction of Magnetization of Basalt in Japan, Tyôsen, and Manchuria." *Proceedings of the Japanese Academy* 5 (1929): 203–5.
3. P. M. S. Blackett, E. Bullard, S. K. Runcorn, *A Symposium on Continental Drift*, Philosophical Transactions of the Royal Society of London, 1965, series A, vol. 258.
4. Ursula B. Marvin, *Continental Drift: The Evolution of a Concept* (Washington, D.C.: Smithsonian Institution Press, 1973), 157.
5. Robert A. Phinney, ed., *The History of the Earth's Crust: A Symposium* (Princeton, N.J.: Princeton University Press, 1968).
6. Bullard explained that "sudden" meant reversing polarity over a period of a few thousand years after oscillating 10 degrees for about a million years.
7. Robert A. Phinney, ed., *The History of the Earth's Crust: A Symposium* (Princeton, N.J.: Princeton University Press, 1968), 233.
8. Ibid., 6.
9. Ibid., 8.
10. W. A. J. M. van Waterschoot van der Gracht, ed., *The Theory of Continental Drift: A Symposium* (New York: American Association of Petroleum Geologists, 1928), 87.

CHAPTER 13

1. Jon Erickson, *Plate Tectonics: Unraveling the Mysteries of the Earth* (New York: Facts on File, 2001); J. J. W. Rogers and M. Santosh, *Continents and Supercontinents* (Oxford: Oxford University Press, 2004).
2. The Greek god Iapetus was the father of Atlas, for whom the Atlantic was named; the Iapetus Ocean was the predecessor of the Atlantic Ocean.

# ❖ Bibliography

Alley, Richard. B. *The Two-Mile Time Machine: Ice Cores, Abrupt Climate Change, and Our Future.* Princeton, N.J.: Princeton University Press, 2000.

Blackett, P. M. S., E. Bullard, and S. K. Runcorn. *A Symposium on Continental Drift.* Philosophical Transactions of the Royal Society of London, series A, vol. 258, 1965.

Cox, A. *Plate Tectonics and Geomagnetic Reversals.* San Francisco: W. H. Freeman, 1973.

Du Toit, Alexander L. *Our Wandering Continents: An Hypothesis of Continental Drifting.* New York: Hafner Publishing, 1937.

Ehrlich, Gretel. *This Cold Heaven.* New York: Vintage Books, 2001.

Erickson, Jon. *Plate Tectonics: Unraveling the Mysteries of the Earth.* New York: Facts on File, 2001.

Erngaard, Erik. *Greenland Then and Now.* Copenhagen: Lademann, 1972.

Fristrup, Børge. *The Greenland Ice Cap.* Copenhagen: Rhodos, 1966.

Georgi, Johannes. "A. Wegener zum 80. Geburtstag." *Polarforschung* 2. Suppl. (1960): 1–102.

Georgi, Johannes. *Im Eis Vergraben.* Leipzig: F. A. Brockhaus, 1955.

———. *Mid-Ice: The Story of the Wegener Expedition to Greenland.* New York: E. P. Dutton, 1935.

Gilbert, G. K. "The Moon's Face: A Study of the Origin of Its Features," *Bulletin of the Philosophical Society* 12 (1893): 262–64.

Hallam, Anthony. *Great Geological Controversies.* London: Oxford University Press, 1989.

———. *A Revolution in the Earth Sciences: From Continental Drift to Plate Tectonics.* London: Oxford University Press, 1973.

Herbert, Marie. *The Snow People.* New York: G. P. Putnam's Sons, 1973.

Holmes, Arthur. "Continental Drift." *Nature* 122 (1928): 431–33.

———. "Radioactivity and Earth Movements." *Transactions of the Geological Society of Glasgow* 18, part 3 (1931): 559–606.

Imbrie, J. K., and K. P. Imbrie. *Ice Ages: Solving the Mystery.* Short Hills, N.J.: Enslow Publishers, 1979.

Kious, W. J., and R. I. Tilling. *This Dynamic Earth: The Story of Plate Tectonics.* U.S. Geological Survey. Washington, D.C.: U.S. Government Printing Office, 1996.

Kirwan, L. P. *A History of Polar Exploration.* New York: W. W. Norton, 1959.

Köppen, W., and A. L. Wegener. *Die Klimate der Geologischen Vorzeit.* Berlin: Bonntraeger, 1924.

Krätz, Otto. *Goethe und die Naturwissenschaften.* Munich: G. D. W. Callwey, 1992.

Kuhn, T. S. *The Structure of Scientific Revolutions.* 3rd ed. Chicago: University of Chicago Press, 1996.

LeGrand, H. E. *Drifting Continents and Shifting Theories.* Cambridge: Cambridge University Press, 1988.

Lindbergh, Anne M. *Listen! The Wind.* New York: Harcourt, Brace, 1938.

Lopez, Barry. *Arctic Dreams.* Toronto: Bantam Books, 1986.

Marvin, Ursula B. *Continental Drift: The Evolution of a Concept.* Washington, D.C.: Smithsonian Institution Press, 1973.

Matuyama, M. "On the Direction of Magnetization of Basalt in Japan, Tyôsen, and Manchuria." *Proceedings of the Japanese Academy* 5 (1929): 203–5.

Oreskes, Naomi, ed. *Plate Tectonics: An Insider's History of the Modern Theory of the Earth.* Boulder, Colo.: Westview Press, 2001.

Pachter, Henry M. *Modern Germany: A Social, Cultural, and Political History.* Boulder, Colo.: Westview Press, 1978.

Phinney, Robert A., ed. *The History of the Earth's Crust: A Symposium.* Princeton, N.J.: Princeton University Press, 1968.

Poncins, Gontran de. *Kabloona.* Alexandria, Va.: Time-Life Books, 1941.

Rogers, J. J. W., and M. Santosh. *Continents and Supercontinents.* Oxford: Oxford University Press, 2004.

Runcorn, S. K., ed. *Continental Drift.* New York: Academic Press, 1962.

Schuchert, Charles, and Carl Dunbar. *A Textbook of Geology.* Part 2, *Historical Geology.* New York: John Wiley and Sons, 1933.

Schulze, Hagen. *Germany: A New History.* Cambridge, Mass.: Harvard University Press, 1988.

Schwarzbach, Martin. *Alfred Wegener: The Father of Continental Drift.* Madison, Wisc.: Science Tech, 1986.

————. *Climates of the Past*. London: D. Van Nostrand, 1963.

Stefansson, Vilhjalmur. *Greenland*. Garden City, N.Y.: Doubleday, Doran, 1942.

Sullivan, Walter. *Continents in Motion: The New Earth Debate*. 2nd ed. New York: American Institute of Physics, 1991.

Totten, Stanley M. "Frank B. Taylor, Plate Tectonics, and Continental Drift." *Journal of Geological Education* 29 (1981): 212–20.

van Waterschoot van der Gracht, W. A. J. M., ed. *The Theory of Continental Drift: A Symposium*. New York: American Association of Petroleum Geologists, 1928.

Vine, F. J. "The Continental Drift Debate." *Nature* 266 (1977): 19–22.

Vine, F. J., and D. H. Mathews. "Magnetic Anomalies over Oceanic Ridges." *Nature* 99 (1963): 947–99.

Wegener, Alfred. *Origin of Continents and Oceans*. Translated from the 4th ed. New York: Dover Publications, 1966.

Wegener, Else. *Alfred Wegener: Tagebücher, Briefe, Erinnerungen*. Wiesbaden: F. A. Brockhaus, 1960.

————, ed. *Alfred Wegeners letzte Grönlandfahrt*. Leipzig: Brockhaus, 1932.

————. *Greenland Journey*. London: Blackie and Son, 1939.

Wegener-Köppen, Else. *Wladimir Köppen: Ein Gelehrtenleben für die Meteorologie*. Stuttgart: Wissenschaftliche Verlagsgesellschaft M. B. H., 1955.

Wilson, Anna B. *Compilation of Various Geologic Time Scales*. U.S. Geological Survey Open-File Report 01-0052. Version 1.0. Washington, D.C.: U.S. Government Printing Office, 2001.

Wilson, J. Tuzo. *Continents Adrift and Continents Aground*. San Francisco: W. H. Freeman, 1976.

Wutzke, Ulrich. *Der Forscher von der Friedrichsgracht*. Leipzig: F. A. Brockhaus, 1988.

# Index